HOW CAN I HELP?

How Can I Help?

Saving Nature *with* Your Yard

DOUGLAS W. TALLAMY

Timber Press · Portland, Oregon

Timber Press
Workman Publishing
Hachette Book Group, Inc.
1290 Avenue of the Americas
New York, New York 10104
timberpress.com

Timber Press is an imprint of Workman Publishing, a division of Hachette Book Group, Inc. The Timber Press name and logo are registered trademarks of Hachette Book Group, Inc.
Printed in China on responsibly sourced paper
Text and jacket design by Adrianna Sutton

The publisher is not responsible for websites (or their content) that are not owned by the publisher.

The Hachette Speakers Bureau provides a wide range of authors for speaking events. To find out more, go to hachettespeakersbureau.com or email hachettespeakers@hbgusa.com.

ISBN 978-1-64326-471-4

A catalog record for this book is available from the Library of Congress.

Contents

Introduction

CHANGE IS AFOOT! The cultural change from an adversarial relationship with nature to a collaborative one is starting to happen, and it gives me hope about the future of biodiversity and thus our own future. In his 1957 book *Wandering Through Winter*, American nature writer Edwin Way Teale noted, "We cannot make the world uninhabitable for other forms of life and have it habitable for ourselves." For the most part, Teale's prescient warning fell on deaf ears—until now! Every day, we are moving closer to making our planet more habitable—not by robbing Peter to pay Paul, but in truly sustainable ways. Like all other species, we humans are evolutionary guests on this planet. Our stay here is provisional: if we destroy our life support systems, our residency on Earth will be a short one. The good news is that we *are* learning how to restore the ecological damage we have wrought in so many places and we are starting to act on that new knowledge. There is momentum in the pace of change, and I have written this book, in part, to try to keep that momentum going.

I see many similarities between the time-sensitive need to improve ecological function on our properties and the late but accelerating race to mitigate climate change. Both the loss of biodiversity and the changing climate are existential crises that demand our immediate attention. Both have been long predicted: We were first warned of the consequences of pumping carbon dioxide into the atmosphere by Swedish scientist Svante Arrhenius in 1896, while sociobiologist E. O. Wilson issued his first of many cautions about the loss of biodiversity in the 1980s. In both cases, warnings were ignored by policymakers until the predicted consequences were no longer predictions, but everyday headlines. The impacts of climate change, including rising sea levels, the acidification of our oceans, melting glaciers, mega-droughts and mega-wildfires, more powerful hurricanes, drying rivers and seas, and more, have received the most attention, and the need to "do something" has finally overpowered self-serving, climate-denying propaganda. Fighting climate change is now accepted as imperative to preserving our way of life.

And so it is with the biodiversity crisis we have created by expelling all but a few species from the landscapes we control. Every day that we continue to banish nature to places we ourselves find undesirable, the crisis grows increasingly dire, propelling us deeper into the Holocene extinction, the sixth mass extinction event on Earth that threatens the life support systems on which we depend. The only real difference I see between the effects of climate change and our defaunation of the planet is not the severity of the consequences of these two threats, but the speed with which they unfold. Though the impacts of species loss are far more subtle and occur far more slowly than a 15-foot storm surge of sea water, in the long run, these impacts will be no less devastating.

Let's develop this comparison a little further. Even though addressing climate change and the biodiversity crisis are mandatory undertakings, there are downsides to both: transitioning from fossil fuels as energy sources will end the centuries old coal mining industry, and jobs will be lost in the near term. It will elevate, at least temporarily, the price of energy and threaten the oil industry that has controlled economies around the world for a century. In the same way, including ecological function as a goal of landscape designs will crimp our reliance on ornamental plant species from other continents and constrain our use of lawns as status symbols. But these downsides are trivial compared to the benefits of switching to clean energy sources and widespread landscaping with ecologically productive native plants. Transitioning to clean energy will mean that people worldwide can live much as they did before climate change but without polluting the atmosphere with climate-altering carbon dioxide. Similarly, abandoning a landscaping culture in which aesthetics always trump ecological functions will create a new ecological landscaping industry; reduce water pollution; diminish flood threats; store more carbon in plants and soils everywhere; sustain larger, more diverse and effective polli-nator communities; restore bird populations and the insects they eat; reduce the likelihood of introducing new pests and diseases from other continents; and dramatically boost sales of native landscape plants.

I like the analogy between fighting climate change and improving ecolog-ical function in our landscapes, but it breaks down when we consider who will implement these solutions. Although individuals can help conserve energy, the big impacts on climate change will come from actions taken by the energy sector, including wind and solar farms, nuclear fusion enterprises, and hot rock and fuel cell industries, as well as bioengineered meat technologies and the like—all specialized industries that require technical training and big money. Changes in our landscaping paradigm, in contrast, can be enacted by everybody and anybody, regardless of our vocation, age, background training, or financial resources. This means that the knowledge required to recognize the need to change our landscaping paradigm and to implement those changes must become far more widespread than it currently is. Based on the number of questions I receive daily on email, we are not there yet, but the interest in acquiring that knowledge definitely is. With hopes of moving the needle toward this end, I have written this book.

Residential landowners, the very people asking the majority of the ques-tions I hear, hold enormous conservation potential because there are so many of them and they control the plantings on landscapes over such large areas of the country. In the United States alone, 135 million acres are now in residen-tial landscapes, and those landscapes are controlled by hundreds of millions of people. With a little education, we all can come to realize that sustainable Earth stewardship is not something we can ignore or practice only if we feel like it, or something that happens only in "natural areas." It is essential, and it is essential everywhere! Moreover, caring for the life around us is a responsibility we *all* hold. I say that with certainty, because each of us depends entirely on the quality of local ecosystems for our continued well-being.

There is one enormous impediment to ecological landscaping that is a byproduct of the culture of private land ownership: the misconception that your property is your discrete and isolated kingdom and you are king. But our properties are not like Las Vegas. You've all heard the expression, "What happens in Vegas, stays in Vegas." Some of you might have put a great deal of faith in that expression at some point in your past. From an ecological perspective, however, what happens in your yard does *not* stay in your yard. Everything we plant, don't plant, or apply to our yard impacts the greater ecosystem in which our yard exists, and thus everything that depends on that ecosystem. For example, when we plant invasive plants, we are harboring ecological tumors whose offspring escape our landscape, negatively impacting our neighbors' properties as well as the natural spaces around us. When we manicure acres of lawn, we diminish the ability of that land to reduce stormwater runoff, compromising the watershed in which we live as well as that of communities tens or hundreds of miles downstream. Planting large swaths of turf grass also wastes an opportunity to fight climate change. Filling our yards with turf grass, notorious for its inability to store carbon, not only does not remove carbon from the atmosphere, but it *adds* carbon every time we mow. Lawn care also pollutes local waterways with the unnecessary fertilizers and pesticides that are routinely applied to grass. When we overuse turf grass, we eschew our responsibility to support the food web that sustains local wildlife. Our plant choices, both in terms of species and quantity, determine how well our landscapes support the diverse communities of native bees required to pollinate the plants that, in turn, are required for high-functioning ecosystems, as well as the insect herbivores that are the bread and butter of terrestrial food webs.

Fortunately, the Las Vegas metaphor works both ways. The way we landscape our yards can just as easily export ecological benefits instead of costs. We can liberally use plants that support the most species of pollinators, both specialists and generalists, as well as the caterpillars, grasshoppers, and other insects that feed our hungry birds. We can reduce our lawns to cover only the areas on which we regularly walk or strips that serve as "cues for care," demonstrating to our neighbors that we are not rejecting the culture of landscape beauty and neatness and are not threatening property values. We can add dense plantings of woody and herbaceous plants that hold rainwater on site and store tons of carbon while doing so, and we can control mosquitoes with benign biocontrol of larvae using mosquito dunks rather than through the ineffective and wide-scale carnage wrought by mosquito fogging.

Invariably, the queries I receive start with something akin to this: "I know you're very busy, but... I have just one quick question." It's not that people don't really believe I am very busy (hardly special these days because *everybody* is very busy), but their eagerness to get answers overwhelms consideration of my time. And that is precisely the kind of interest I am trying to encourage, so I answer as many questions as I can. This book is a compilation of those questions and answers, aggregated by subject into chapters. I have favored questions about general ecology and evolution as well as the biodiversity crisis because some understanding of these subjects is required to grasp why we need to change the way we landscape. Questions about why, how, and to what extent native plants outperform non-native species are also treated at length because the switch from favoring non-native ornamentals to productive native plants is the biggest change we need to make in our landscapes to mitigate defaunation.

My book *The Nature of Oaks* has stimulated much interest in these great trees but also many questions about oaks. I have also targeted questions about invasive species, including plants, animals, and pests such as ticks and white-tailed deer—all of which make ecological landscaping harder than it should be. Many people have asked important questions about conservation, restoration, and the effort to create a network of homegrown national parks across the country. And it's no surprise that I receive questions about the sociological aspects of ecological landscaping: How can we reduce lawn, keep our leaves in fall, and use many more native plants without running afoul of township or HOA regulations? Finally, I have grouped specific questions about supporting wildlife at home, including the beloved monarch, songbirds, pollinators, and the unsung heroes of terrestrial food webs, caterpillars.

Early questions in each chapter are more entry level or generalized, with more specific questions following under appropriate subheadings. Questions have been reproduced verbatim or with slight edits for brevity or clarity. Questioners remain anonymous for obvious reasons. I often get questions asking for specific plant recommendations; for the sake of brevity and because such answers are easily found on rapidly growing web resources such as the National Wildlife Federation's Native Plant Finder, California Native Plant Society's Calscape, the Wild Ones website, and many others, I have not included those questions. Although you are welcome to look up the answer to your specific question and throw the rest of the book away (on second thought, donate it to your local library), my hope is that you will read my answers to all the questions, as many are interrelated and build on answers that have come before. I have tried to remain apolitical throughout the book, which has been easier than you might think because everyone—regardless of their political views, where they live, their age, or their socioeconomic background—requires a healthy environment and productive ecosystems. No exceptions.

One of the most common questions I'm asked in interviews is about how I became interested in entomology and teaching. I'm not sure why that matters, but I'll answer it here anyway. My love of nature is not my fault: I was born loving nature, or at least the animal part of nature, and had no say in the matter. Toy cars and trucks could not hold my attention for long. But those plastic dinosaurs I got out of cereal boxes and, later on, real snakes and turtles—now there was fascination. Plants didn't do much for me, although I did like big trees and the forests they created because they were the homes of the animals I loved. Forests also promised adventure. Maybe I could get lost in an endless forest and have to build a log house to survive winter. I could fashion wooden spears out of saplings, just like Daniel Boone did as a boy, and live off the fish I would spear in streams. I could even start fires by banging two stones together. I never had much luck with that, but I did make some pretty good spears.

I also was born wondering how nature worked. I remember standing in my driveway in Plainfield, New Jersey, when I was five years old, wondering how wind got started. In a stroke of pure genius, I figured it out: Someone on the other side of the world jumped. That rustled a leaf, which then rustled two leaves, which then got a whole branch of leaves moving, and before you knew it, there is wind moving all over the planet! Another mystery solved!

Insects entered my life in the usual way: I took a course in college. I knew nothing about insects before I went to college, but in my junior year I took a course in invertebrate zoology and learned about the various orders of insects. That was so interesting that I signed up for the only course in entomology that was offered, which was taught by Dr. Robert E. Bugbee.

Dr. Bugbee, as you might expect, knew his insects and had no trouble at all in shifting my interest from snakes and turtles to those wonderful six-legged creatures that, as E. O. Wilson put it, "run the world."

How I started teaching is best described as reluctantly. I may have been born loving nature, but I was also born utterly terrified of public speaking. In seventh grade, we had to give a two-minute speech to the class. TWO WHOLE MINUTES! Who can talk for that long? It was a heart-stopping experience that convinced me that I would never, under any circumstances, subject myself to it again. Even after I graduated from college, I was a contender for the world's worst public speaker. I would breathe in while talking, but then forget to breathe out. In no time, my lungs would fill and I would start to choke. Needless to say, I scratched teaching off the list of career possibilities very early on. But I did need a job with a paycheck to help support my young family, so I applied for research positions at various universities that had a teaching component. In 1981, I started my job at the University of Delaware, where I immediately had to teach two courses that I created from scratch. I think I survived because I spent so much time frantically writing the lectures (I was usually just one lecture ahead of the syllabus) that I didn't have any time or energy left to fret about delivering those lectures. Over the years, I learned an important lesson in life: you can get used to anything. Before too long, the students in my classes were the same age as my own kids, so lecturing to them was not nearly as scary as lecturing to the professors who had written my text books, something that had often happened at scientific meetings. And today, talking in front of people is a non-event for me. When you stop worrying about lecturing itself, you can focus on what you are actually saying, and then it becomes just like talking to a friend.

How did I come by the knowledge and experience I offer in the coming pages? I took my formal academic training in biology, entomology, and ecology at Allegheny College, Rutgers University, University of Maryland, and University of Iowa. My informal education about restoring ecosystems at home began when my wife, Cindy, and I bought a 10-acre section of a defunct 300-year-old farm in southeast Pennsylvania. The land was a mess. The last farming that had occurred on the property was mowing for hay, and mature trees were scarce. By the time we moved in, the property had been thoroughly invaded by Asian plants such as multiflora rose, oriental bittersweet, autumn olive, Callery pear, Japanese and Amur honeysuckle, privet, and many more non-natives. It was during my early experiences on this degraded piece of land that I first realized non-native plants are poor at supporting native insects, that non-native ornamental plants are favored over native species in residential landscapes nearly everywhere, and that many of these ornamentals are the serious invasives that have penetrated natural areas throughout North America, degrading food webs and pollinator communities wherever they have spread. I also discovered that ecologists at that time did not consider human-dominated landscapes as viable options for conservation, so there had been little research documenting these problems or seeking solutions to them. In other words, the early walks on our new property redirected my research from asking how cucumber beetles choose their mates to asking how we can mobilize property owners everywhere to become a powerful force in conservation.

I have discovered many things during my switch from heuristic research in ecology to investigating practical issues in restoration biology. Perhaps the most important is that Mother Nature is far more resilient than most people realize and that, given half a chance, she will repair much of the damage we have inflicted on her in short order. Cindy and I have witnessed this first-hand. Up to now, 1323 species of moths and 62 species of birds have bred on our property (the only two groups I have counted so far). We didn't take the time to count the moths and birds making a living on our property when we first moved in (I wish we had), but I can assure you that it was far, far fewer. The return of life to our land has happened in just a few years for one simple reason: we put the plants back—or at least some of the plant species that used to call our lot home. Our success makes me wonder what would happen if everyone put the plants back.

I have also discovered that most people want to help fight the biodiversity crisis, but they feel powerless to do so. What can one person do? In the coming pages, you will learn what one person can do. You will learn that by reducing the area of lawn, favoring the use of keystone native plants, replacing white bulbs with yellow bulbs in outdoor fixtures, firing your mosquito fogger, planting for specialist pollinators, retaining the leaves that fall from your trees each year, removing the invasive plants on your property, and much more, you can totally revitalize ecosystem function on your property, enhance your surrounding ecosystem instead of continuing to degrade it, and watch these changes happen before your eyes. In short, you will become empowered; you will join millions of other like-minded people to become the future of conservation. Pretty neat!

A few of my answers reflect my own opinions based on years of experience with what works and doesn't work. And when my answer is just my opinion, I say so. But nearly all of my answers reflect our best current knowledge honed over the years by the scientific method. It's important to remember that science is a process—the process of hypothesis testing. It is *not* someone's opinion or belief. A scientist asks a question, restates it as a hypothesis, designs an experiment to test that hypothesis, and accepts or rejects the hypothesis based on the results of that experiment. Scientific consensus can change as more and more experiments are performed and more data are generated, and I realize this frustrates the public. It seems like scientists can't make up their minds. Just tell us what is right and what is wrong! Is coffee good for you bad for you? Black or white? I wish life were that simple, but most of the time, the answer lies somewhere in the gray area. It is my hope that this book will provide enough background knowledge and nuance about landscape restoration to remove much of that frustration.

CHAPTER ONE

Ecology and Evolution

Can you briefly explain what evolution is and how it works?

The formal definition of evolution is very short: a change in gene frequency over time. As time goes on, genes within a population of an organism change in how common they are. These changes are usually (but not always) the result of natural selection, the differential survival and reproduction of individuals. Individuals carrying genes that enable them to survive and reproduce better than other individuals tend to pass on more of these genes to future generations than individuals carrying genes that make them less fit. If having blond hair, for example, enables blondes to survive and reproduce better than brunettes, there will be more blondes in the future than brunettes. That's all there is to it. But it is important to remember that evolution does not proceed through binary choices among selection pressures; there are usually many selection pressures, or evolutionary forces, that shape a trait. Let's use pottery as an analogy. We typically envision a single person shaping the clay to their desired goal. But if evolution were writing this script, dozens of potters would all have a hand in shaping the clay, and the end result would be a grand compromise based on the strength of each potter's influence. In other words, evolution is a compromise that occurs as a result of competing selection pressures.

I hear people talking about something being good or bad for our ecology. Is that the proper use of the word *ecology*?

No, it is not. Ecology is the study of how living things interact with one another and their physical environment. When people talk about "our ecology," they are incorrectly synonymizing *ecology* with *ecosystem*. An ecosystem is a biological community of interacting organisms and their environment. All things live within an ecosystem, but nothing has "an ecology."

I've heard the terms *ecological source*, *ecological sinks*, and *ecological traps*. Can you explain the difference to me?

An *ecological source* is a population that is large enough to produce individuals that disperse to other locations without seriously reducing the number of individuals in the population. In US history, for example, the East Coast served as a source for colonists who moved across the country without depleting the East Coast's human population. An *ecological sink* is a habitat that attracts members of a species but is unable to sustain them. Thus, in the long run, sinks deplete overall numbers because all individuals that end up in a sink die. This is where the difference between a sink and a trap becomes very fuzzy. When a habitat produces cues (favorable environmental/chemical attributes or events) that attract colonizers of a particular species, but there are insufficient or unsuitable resources to sustain those colonizers, that habitat becomes an *ecological trap*. Consider, for example, an oak tree in a suburban yard that is not properly landscaped. The chemical perfumes (volatiles) produced by that oak tree will attract a number of species of moths whose larvae develop on oaks. Those moths will lay eggs, the caterpillars will grow and reach the point where they must pupate, and then they will drop from the tree and try to burrow underground to spend winter as pupae. If the ground under the tree is mowed and compacted, a condition present under most yard trees in the United States, the caterpillars may not be able to burrow underground and will die without reaching adulthood. That oak tree, then, becomes an ecological trap. In short, when an animal mistakenly prefers a habitat that actually lowers its ability to survive and/or reproduce (its fitness), that habitat is acting as an ecological trap.

What are biological corridors, and what happens when we've lost them to roads/highways, housing developments, shopping centers, and the like?

What you are really asking is, what is the problem with small or tiny populations? Because that is what happens to large populations when we fragment and isolate them with roads, developments, and other human-made infrastructures. Two main problems arise when a population is reduced in size.

The first problem, arguably the most serious one, is that small or tiny populations are highly vulnerable to environmental perturbations and local extinctions for several reasons, the most important being that all populations, whether they are tiny or large, fluctuate: in good times they grow and in bad times they shrink. Let's say, for example, that a population experiences a period of benign weather that reduces physical stress on its members and also creates an abundance of food. These favorable factors lead to more successful reproduction than usual, so the population grows faster than predators can exploit it. Then the weather turns unusually cold or dry, food becomes scarce, natural enemies catch up with their prey, and deaths in our hypothetical population exceed births. The result is a natural decline in numbers that will continue until conditions improve. These cycles in population size are normal and occur over time to a greater or lesser extent in every species on the planet. When a population is large, it typically can withstand even severe environmental insults, but when it is reduced to a small fraction of its normal size, environmental stress can send it into oblivion.

A second problem that arises in small populations is inbreeding. Imagine a box turtle isolated within an urban woodlot with nine other box turtles. Five turtles are males and five are females. There is plenty to eat, so that's not a problem. But over the next several generations, there are so few mate choices that it becomes almost impossible for each turtle to find a mate that is not a close relative. The worst-case scenario is brother mating with sister. This is called inbreeding depression, and it doesn't take long for genetic abnormalities to appear and spread through small populations that do not benefit from an influx of genetic variability on a regular basis.

As a solution to the threat of local extinction and inbreeding depression, ecologists tell us we must reestablish connectivity among small, isolated populations—that is, we need to provide biological corridors, or strips of habitat that enable plants and animals to move freely from one habitat fragment to another. If habitats are connected by biological corridors, they are no longer truly isolated, so the populations they house will no longer be perilously small. Biological corridors are also a means by which individuals with new genes can enter populations, preventing issues from inbreeding depression. I agree with this solution but would like to embellish it a bit. Rather than just building skinny biological corridors among viable habitats, why not restore the viability of as much of the space between habitats as possible? We can do this to a greater or lesser extent by restoring native plants that are the foundation of sustainable populations everywhere, not just in narrow strips connecting viable habitats. In other words, if we restore viable habitats in many of the places in which we have destroyed them, populations will no longer be small or isolated in tiny habitat fragments.

Why should I care about the food web? If we lost most of the caterpillars, how would that affect me—and other humans?

Quite simply, the food web transfers energy that plants harness from the sun to other organisms. Some animals get their energy directly from plants, while others get energy indirectly from plants by eating something that ate plants or by eating something that ate something that took energy from plants. This transfer of energy from plants, to plant eaters, to predators—that is, across multiple trophic levels—is what enables a vast diversity of animal species to exist on Earth. So when you ask why you should care about the food web, you are actually asking, why should I care whether complex life can continue to exist on Earth? No food web, no animals. No animals, few to no flowering plants. No flowering plants, vastly simplified food webs and the loss of most species on Earth.

Why can't we humans be the only animals on the planet, maybe along with some cows, pigs, and chickens? Because it takes highly diverse ecosystems to produce the life support we *all* depend on every day. The oxygen, weather stabilization, flood control, carbon sequestration, water purification, topsoil creation, wood products, and so much more produced by plants would not be here without species-rich communities of pollinators, pest control services provided by myriad animal species, seed dispersal supplied by birds and mammals, nutrient recycling provided by insect decomposers, and other factors. Without high-functioning ecosystems, humans would last only a few months, and there would be no high-functioning ecosystems without animals and the food webs that support them.

What are your thoughts on novel ecosystems that are beginning to appear in disturbed places (whether by design or not) that are biodiverse and complex without necessarily reflecting historical ecosystems?

My goal has never been to restore ecosystems to what they were at some point in the past. We have changed too much locally and on such a grand scale that we can never achieve that goal. But we *can* restore at least *some* of the interactions among animals and plants that historically occurred at a site to improve the stability and productivity of that ecosystem today. For me, that is a worthy enough goal. Besides, most ecosystems are quite dynamic and have achieved resilience through change in the past. The specialized relationships among plants and animals are far more stable and can last millions of years, even if they shift in space over time. This is why evolutionarily novel ecosystems are less productive than coevolved ecosystems—they do not contain the specialized interactions between plants and animals that form over millennia.

You often say that human overpopulation is the root of our environmental problems. If that is so, why does Elon Musk call population decline the greatest threat to humanity? And why is the news that China is no longer growing in population being called a crisis?

The short answer is that Elon Musk is as ignorant of the ecological forces that drive life on Earth as he is knowledgeable in other ways. Recent headlines about population declines in China, Japan, Russia, South Korea, and Italy have indeed been met with cries of alarm. The fear has been explicit: fewer babies will create an existential crisis for humanity. One reason for the emergency—the one we unfailingly hear about first—is that the decline in birth rates means there will be more old people than young people for at least a few decades. Assuming the old depend on the young for care, there won't be enough young people to take care of all of the old people.

But it is not humanitarian challenges that strike fear in the hearts of economists when population growth slows. Rather, population declines threaten age-old economic models that assume perpetual growth. Fewer people equals fewer consumers. Our measures of "success" tend to be economic in nature, and they assume that we can—indeed, we must—enlarge our economies forever. If our GDP does not increase annually, if we don't consume more and more goods, most economists tell us we are doomed. The concept of a steady-state economy—one that maintains productivity at an even keel without growing, an economy in which we do not constantly increase our material wealth but that produces enough for all—is anathema to dominant economic models.

And so, we are told, we have a crisis, and many governments are scrambling to incentivize baby-making. What we have *not* been told is that reports of population declines are the first glimmer of real hope that we humans may have a future on Earth. No matter how many times or how cleverly we use the word *sustainable*, nothing on this planet is sustainable if the human population continues to grow. We *do* face existential crises today—several of them, in fact: We're in the midst of the sixth mass extinction event on Earth, which is rapidly eliminating the organisms that run the ecosystems on which we all depend. Climate change is causing extreme droughts, floods, wildfires, sea level rise, heat, and ocean acidification. Abundant fresh water is disappearing where we need it, rivers do not reach the sea, and immense aquifers are nearly dry. Global pandemics are inevitably plaguing crowded populations. These are real crises, all of which are caused by too *many* people, not too few.

And yet, incredibly, we decry the first good news about human population growth as a crisis! By "we" I mean primarily economists. Every card-carrying ecologist on the planet agrees that our global population has exceeded Earth's carrying capacity—its ability to sustain human numbers without degrading the resources we and other living things require in the future. In fact, most agree it would take from two to four Earths to sustain our current population at a reasonable level of comfort. On a finite planet, perpetual growth is simply not an option. We have no choice but to transition to a no-growth culture as soon as possible. Transitions are often hard, though, because we are forced to find new ways to deal with old problems. Curiously, the challenge we hear about more than any other—the problem of what to do about old people—may actually be the easiest to meet.

We do have old age/retirement facilities, but, as they are currently run, they will not meet the needs of large aging populations for two reasons: these facilities are too expensive for most people, and too few of the facilities exist, even if they are affordable. The second point is easy to fix: build more facilities. But what about the first? Who will care for old folks and how can we make that care affordable? A recent headline suggested that robots will save the day. Robots aside, consider this: What if we restructured care facilities so that old people take care of other old people? Fortunately, we all don't grow old and incapacitated at the same time. Millions of people remain healthy and productive for 20 or more years after retirement before they themselves need care. They are, in fact, an unlimited labor force. Who will pay these elder caregivers? No one. Instead, we can establish a barter system: If you help care for your peers in our facility, we will provide you with free room and board. Bingo! The need for an enormous young workforce to care for the elderly as well as the exorbitant price tag associated with these facilities disappears. The vast majority of care old folks require is not skilled medical attention, but unskilled efforts, ranging from moving people about in wheelchairs to helping with meals (cutting up vegetables for a salad can be impossible for many oldsters), or even simple companionship.

I realize I am being optimistically simplistic with this solution, and many experts will be eager to tell me why it won't work. For example, there must be some income stream into care facilities for them to operate—where will that come from? Yet ideas like this are points of departure meant to get us thinking about real solutions instead of pretending the problems don't exist in the first place. Our real crisis has been, and continues to be, reliance on an economic Ponzi scheme that is designed for infinite growth on a finite planet. If we don't accept the limits of our planet soon, this will end as all Ponzi schemes end. But rather than bankrupting only the people who were duped into investing in the scheme, the myth of perpetual growth will empty our ecological bank account—the one that, when healthy, provides everyone with our day-to-day needs. It is past time that we confront the problems posed by overpopulation, celebrate the recent news of growth declines, and use our enormous brains to find ways to live sustainably on the tiny blue marble that is planet Earth. There truly is no Planet B.

Why is caterpillar diversity in your yard a good thing?

This question goes hand in hand with the oft asked question, why is species diversity touted as a good thing? A typical ecology textbook would say that diversity is good because it increases ecosystem productivity and stability. But what does that mean?

Let's say a pair of Carolina Chickadees wants to breed and nest in my yard. They need two things to do this successfully: a cavity supplied by a tree hole or nest box and food, largely in the form of easily digested caterpillars rich in protein, fats, and carotenoids. And not just a few caterpillars—they need 350 to 570 caterpillars every day (depending on the number of chicks in the nest) to keep their nestlings well fed. That adds up to 6000 to 9000 caterpillars required to feed the chicks until they have fledged. But for another 21 days after the nestlings fledge (we don't know exactly how many days because they fly and it's impossible to count them accurately), chickadee parents continue to feed them caterpillars. Suffice it to say that it easily takes more than 10,000 caterpillars to produce *one* nest of independent chickadees—an astounding number when you think about it.

Now, if I want this chickadee pair to raise its young in my yard, all of those caterpillars will need to be in my yard because, like most birds, chickadees forage very close to the nest—on average, 50 meters (about 164 feet). They are not flying five miles down the road to the nearest woodlot to forage. And here's where the importance of diversity comes in. Insect populations fluctuate widely from year to year, and if only a few species of caterpillars were in my yard and it happened to be a bad year for those species, there wouldn't be nearly enough caterpillars to meet the needs of my chickadee family. If, however, many species of caterpillars were living in my yard, it is unlikely that they would all be in a down cycle at the same time, so there would always be enough caterpillars available to enable the chickadees to breed successfully every year. And I *do* have many caterpillar species in my

Cucullia pulla is among the 14,000 or so species of caterpillars in North America.

yard—in fact, I've recorded 1259 moth species and I'm still finding new ones on a regular basis!

The diversity of caterpillar species in my yard has created redundancy in my ecosystem: several species are doing the same job. If one species' numbers are low in a particular year due to weather or an abundance of specialized parasitoids, other species are present to compensate. Diversity also increases the chances that one or more species will have life histories that make them common (abundant). This is important because most species in a community are relatively rare. Just a few, the keystone species, do most of the ecological work. As diversity increases, so does the chance that a keystone species will join the community. Because of redundancy and the improved chance of housing a high-producer, diversity enables my chickadee pair to breed every year, not just when all the stars align. We call that ecosystem stability, and it is the direct result of the diversity of caterpillars in my yard. Needless to say, I have a diversity of caterpillars only because my yard includes a diversity of the host plants those caterpillar species need to develop.

Plant Ecology

Plants evolved way before insects and birds. How did they manage without pollinators and seed dispersers?

The first land plants to evolve were bryophytes, algae-like nonvascular things, about 500 million years ago. Then came mosses, followed by ferns, both of which reproduced by spores and did not require pollination or seed dispersal. Gymnosperms, like present-day conifers, evolved seeds next. These plants required pollination and dispersal, but the wind accomplished both of these ecological jobs. It wasn't until angiosperms, the flowering plants, evolved about 275 million years ago that insects and plants developed early mutualisms to achieve pollination. By then, there were plenty of fruit-eating reptiles around to disperse plant seeds. True birds did not appear until 60 million years ago, but when they did, they kicked seed dispersal into high gear.

How do plants grow in the shade?

Several things are essential to plant growth, but if it's possible for one thing to be more essential than anything else (though by definition, that's not possible), then light would be that thing. All plants need sunlight because energy from the sun is converted through the miraculous chemical reactions of photosynthesis into the sugars and carbohydrates required by plants to grow. Light, however, can be in short supply under certain conditions, so plants often must compete for it. In a forest, the winners of this competition are the tall canopy trees. Indeed, the need to harness as much light energy as possible has undoubtedly been the primary selective force favoring the great heights achieved by so many tree species. But tree canopies do not gobble all of the light, and this creates niches where plants with lesser light requirements can grow below canopy species. This is why a typical forest has several horizontal layers. The canopy is the top layer, and just beneath the canopy is the subcanopy layer, which usually comprises younger canopy tree species that are waiting for the demise of a nearby mature tree. Beneath subcanopy trees are understory trees such as dogwoods, ironwoods, and witch-hazels that are adapted to make do with lower light intensities. Lower still are shrubs such as blueberries and azaleas and, finally, hundreds of plant species that grow close to the ground where, during periods of full leaf expansion, there is very little light at all.

So how do plants that occur in low light thrive? They use several strategies, some of which are nicely displayed within single trees. Consider the leaves on a nearby oak. Leaves on the lower branches, where much of the light has been blocked above, are larger, sometimes several times larger, than leaves in full sun at the top of the tree. They are also less lobed with more shallow incisions between lobes. These adaptations increase the surface area of individual leaves so they can gather more light. Many of the plants in the ground layer are spring ephemerals, meaning they grow quickly in spring before leaves in the layers above them block light. Some, like Virginia bluebells, spring beauties, and bluets, grow, flower, seed, and die before much light is blocked, while others, such as mayapples and bloodroot, grow and flower quickly but remain alive throughout the growing season, using their broad, spreading leaves to harvest the light energy that reaches them. The products resulting from these reduced levels of photosynthesis are then stored in large tuberlike roots for use the following year. Some plants, such as pink lady's slippers, can live under low light conditions without flowering for decades. Only when a treefall or forest fire creates a light gap will they have the energy they need to flower and produce seed.

Please explain the root structures of trees. How far do they extend? How deep are they? Are they different in a woodland versus a yard? Where are the nutrients and water being absorbed—near the trunk or out beyond the dripline?

Roots extend much farther from the trunk and not as deep in the ground as most people think. The standard lore has been that tree roots stop at the dripline of the canopy. This perception is reinforced whenever we see a blowdown, because the roots of the downed tree don't extend very far at all. But if you look closely, you can see that all of the roots have been broken (or rotted) off in a blowdown. The small, fine roots that deliver nutrients and water to the tree are actually still in the ground. We now know that the roots of many tree species extend at least twice the distance from the trunk to the dripline. (One study group in Europe dug up the roots of a large English oak—or at least they tried to. Some 300 feet away from the trunk, they gave up!)

We often hear that trees such as oaks and hickories have deep tap roots, and we envision a root that grows straight down for dozens of feet on a mature tree. Some tree species do have relatively deep tap roots when they are young, and that makes them difficult to transplant. But as the tree gains size, it sends its roots laterally. Remember what roots are for: they support the tree, yes, but their primary role is to the absorb water and nutrients needed for growth. Thus, tree roots grow where the soil contains water and nutrients. Nitrogen, phosphorus, and other nutrients are most plentiful in the upper regions of the soil where decaying plant parts have deposited them. In areas with little rain, tree roots can take a deep dive to find water, but in much of the country east of the Mississippi, the top horizons of the soil contain plenty of water that is replenished every time it rains. Both nutrients and water are absorbed at root tips that grow farthest from the trunk through fine root hairs, often in association with mycorrhizal fungi.

Can trees talk to one another?

They can, but not in the sense that we talk to one another. Trees and other plants can send, receive, and respond to signals—that is, communicate with one another—and they do so in two different ways. The first is through airborne chemicals. When a leaf is injured, say by an insect, it releases chemicals into the air at the site of the injury. You can smell such chemicals every time you mow the lawn. Nearby plants also detect the chemicals in the air and, in response, produce defensive chemicals that protect their own leaves. The more closely a plant is related to the injured plant, the better it is at detecting and responding to these chemical signals. Plants also communicate below the ground. An injured plant can release chemicals into the soil, which are then transferred to other plants via mycorrhizal fungi. Or the roots of one plant can actually graft onto the roots of another plant and exchange nutrients—as well as plant diseases, unfortunately. Suzanne Simard, a forest ecologist, has conducted research that suggests there is a great deal of underground communication and nutrient exchange, at least in the Douglas fir forests of the Pacific Northwest.

A few plants support many species of caterpillars. But are the plants that don't still valuable?

Yes, indeed they are. Diversity itself is a good thing, and studies have shown that as the number of species in an ecosystem increases, so do ecosystem productivity and stability. But plants that support fewer caterpillar species still contribute to the productivity and stability of their ecosystem. Moreover, nearly all plants support at least a few species of caterpillars, many of which are specialists on those plants. Without the plants, we would lose these caterpillars. Tulip poplars, for example, support only 21 species of caterpillars in the mid-Atlantic States, 26 times fewer species than supported by oaks. But without tulip poplars, we would not have the tulip tree silk moth. Greenbriars support only 19 caterpillar species, but without these plants, there could be no curve-lined owlet, spotted phosphila, or turbulent phosphila moths. When the contributions of the plants that do not play keystone roles in the food web are combined, they become as substantial as those of many keystone plants. The best landscape will have a diversity of plants in addition to the keystone powerhouse species, all of which add to the redundancy in a yard's food web.

The beautiful curve-lined owlet caterpillar depends entirely on greenbriar.

When I plant a new tree, do I need to inoculate the soil with mycorrhizae?

The short answer: probably not. Mycorrhizal spores float on the wind pretty much everywhere and will eventually colonize the soil under your tree. But you can speed up the colonization of the soil with the appropriate mycorrhizae species by adding some soil (humus, really) dug from under a mature tree of the same species to the soil under your new tree. It is best if the host tree is in a natural setting, and it doesn't have to be much humus. The mycorrhizae in the inoculant you transfer will grow quickly and colonize all the rootlets of your new planting.

You have stated that in the eastern United States, most meadows and prairies are temporary, unstable ecosystems because they receive so much rainfall that without continued disturbance, such as grazing or fire, they will move through succession to forests. Conversely, most wildflowers discussed as food for bees, for monarch way stations, and so on, are plants that need full sun. Are they not stable and self-sustaining?

In the long term, meadows and prairies are not stable. They will be overgrown by woody plants unless the woodies are kept at bay. Remember that part about disturbance in the form of fire or grazing? Before humans inhabited North America, huge numbers of very large mammals, many of which were browsers (they ate woody branch tips, particularly in winter), kept woody plants out of sunny meadows. Beavers the size of Volkswagens, giant sloths that could reach up 18 feet to browse, wood bison, mammoths, several species of rhinoceroses, camels, horses, elk, deer, and many others all kept much of what is now closed canopy, dark forest open in a more savanna-like landscape. The result of this disturbance was that prairie plants were plentiful, even in the eastern parts of North America. When humans did arrive, they managed eastern ecosystems with fire, which also allowed more sunlight into our forests than we typically see today.

How can I tell the age of a tree?

In the temperate zone, where trees grow only during the warm seasons, each year of growth for a tree is recorded with a distinct ring laid down by the cambium. These rings are clearly visible in the woody xylem if the tree is cut down. Aging trees without cutting them down requires an increment borer, a tool used to bore into the bark of the tree to its center core. It produces a slim cylinder of wood; all of the tree rings are visible along this cylinder. But the easiest way to age a tree is to learn about the history of the land around it. For example, if you want to know the age of trees on your property, find out when your house was built and what was on the land before it was built. If it was farmland, then you will know that the trees in your front yard are probably no older than your house. You may be surprised at how fast the trees have grown.

What is your position on genetically modified agriculture?

I have tried to keep my work as apolitical as possible. Everybody requires healthy ecosystems regardless of their politics. Genetically modified agriculture is a scientific issue, not a political one. It is a tool in our pest control toolbox. Its use should be compared to wide-scale use of pesticides, often sprayed from planes, when measuring its impacts. Can it be abused? Yes. Has it been abused? Yes. Is it always abused? No. Can it reduce overall use of pesticides? Yes. Is there a danger of genetically modified genes escaping into the environment? Almost certainly no. Genes move among plants through pollination. There is little danger of corn pollen fertilizing nearby trees, asters, or anything but a close corn relative, and there are no close corn relatives in North America. In short, fear of GMO technology is based more on misinformation than on empirical facts.

Animal Ecology

Why can't insects adapt to imported plants that have been here for hundreds of years? Studies have shown rapid adaptation in Darwin's finches and fruit flies, so it seems possible.

This is a reasonable question. In fact, this is one of the most important questions in this book! The answer is multifaceted. First, there is no pressure for insects to adapt to a new host. They have good hosts already, and they look for those instead of trying to use new ones. Darwin's finches had no options—they had to eat new seeds or die. So there was very strong selection for slight variations in beak morphology that enabled birds to exploit new seed resources. Our insects don't know we have removed most of their host plants, so they instinctively search for the chemical cues they have used for millions of years.

The second reason it is very difficult for insects to adapt to new host plants is the number of genetic changes that would be required to do so. Host plant specialization involves hundreds of genes—genes that control where the adult female searches for host plants, what volatile chemical cues from those plants she will be able to recognize, what surface waxes on the plant stimulate her to lay eggs on that plant, whether the hatching caterpillars are stimulated chemically to start eating the plant, and genes that enable the caterpillar to handle the chemical defense of that plant physiologically. This is far more complicated than a simple enzyme change that enables an insect to survive a toxic spray. Think of all the genes that would be necessary to direct an adult female to recognize a new plant as a potential host, to lay her eggs on it, and then to have the larvae physiologically able to eat the plant without dying—and all of these mutations happening at once in enough individuals so there would be a breeding population able to use a new plant. The chances of that happening are slim to none.

The monarch butterfly's host relationship with black swallowwort is a good example of how difficult host switches are in nature. Black swallowwort (*Cynanchum louiseae*) was introduced to the United States from Europe as an ornamental in 1853. Before long, it escaped our gardens and is now a serious invasive species throughout much of New England. Black swallowwort is a member of the milkweed family, and the sensors on the monarch's antennae, tarsi, and ovipositor recognize it as a milkweed. Unfortunately, this leads monarchs to mistake black swallowwort for their usual milkweed host and lay eggs on the plant, which is a serious problem for monarchs, because swallowwort's defensive chemistry is different from what is found in the genus *Asclepius*, and monarch larvae do not have the enzymes they would need to detoxify it. And so, every tiny hatchling that finds itself on black swallowwort because its mother mistook the plant for a true milkweed starves to death. Even after 170 years of such mistaken exposures, monarchs have not developed sufficient adaptations to be able to develop on black swallowwort. Host switches do happen in nature, but they are extraordinarily rare.

You claim it takes thousands of years, if ever, for insects to adapt to new host plants. But some other entomologists disagree and point to several species of butterflies that now breed successfully on non-native plants. How do you explain that?

When I say it take eons for insects to adapt to a new host, I am talking about host *switching*, not host *range expansion*. In host switching, new adaptations are required by the insect to adapt to plants in evolutionarily novel lineages with phytochemical defenses the insect has never encountered in its evolutionary history. Host switching happens, but rarely, for all of the reasons I describe in the previous answer. Suppose, for example, that monarchs all of a sudden added snapdragons to their larval host repertoire. It is theoretically possible that this could happen, but I am not holding my breath. Snapdragons deploy totally different chemical defenses in their leaves, and it would take substantial evolutionary change for monarch caterpillars to become able to eat snapdragon leaves without dying. It would also take a considerable change in adult monarch sensory organs for females to decide to lay eggs on snapdragons. Changes of these magnitudes are very rare indeed.

Host range expansion, in contrast, occurs when an insect starts using a new host plant for which it already has developed the necessary adaptations. The black swallowtail, for example, is a specialist on plants in the carrot family, and it already has the adaptations required to exploit species in the carrot family. When we brought Queen Anne's lace, carrots, parsley, and dill to this country, the black swallowtail already had the adaptations required to eat those plants. So it started using them as well as native hosts like golden Alexander. In this way, it expanded the range of plants it uses successfully for larval development. Host range expansion can happen as soon as the insect and new plant meet, because no further evolution is required. Most of the examples people offer of native insects readily using non-native plants are cases of host range expansions, not true host switches. But even accounting for host range expansion, the vast majority of our native Lepidoptera cannot use non-native plants. In fact, my research assistant Kimberley Shropshire and I have found that 15 times more Lepidoptera species use native ornamental plants than non-native ornamentals.

Are caterpillar host records recorded at the plant genus level a good index for the caterpillar productivity of all of the species in a genus?

First, let's look at why we record host records at the genus level. Host plant records are scattered throughout the literature, and many of them are quite old, sometimes reaching as far back as the late 1800s. It was (and still is) commonplace for natural historians and scientists to record host plants at the genus level. Comments like "eats oaks" or "eats maples" are far more common than "eats white oak" or "eats red maple." If a caterpillar is commonly collected on white oak, red oak, pin oak, shingle oak, post oak, bur oak, and so on, it is far easier to say "eats oaks" than to list every oak species it has been found on. Specific host records do occur in the literature, but they are far less common; if we used only these in creating our productivity index, we would lose more than half of the host records that actually exist.

Are there serious trade-offs when we record hosts only at the genus level? One possibility is that summing all the host records over the entire genus in a particular region inflates the number of caterpillar species any one member of that genus can support. The other possibility, and the assumption we have made, is that if a caterpillar is adapted to eating one member of a genus, it will have the adaptations required to eat all members of the genus. There are good reasons to make this assumption. Plants defend themselves against caterpillars with phytochemicals that are usually traits of every species in a genus. For example, oaks protect themselves with tannins. All oaks have tannins, so if a caterpillar species discovers through evolutionary time how to counter tannins as a defense, it would then be able to eat any oak that relies on tannins for defense. I am sure the number of caterpillar species supported varies across species in a genus, but the variation may be minor compared to the variance among genera in records of caterpillar use.

For example, oaks support 557 species of caterpillars in the mid-Atlantic region. Tulip poplars support only 21. You might argue that there are 20 species of oaks in the mid-Atlantic region and only a single species of tulip tree, and that any of the 20 oak species may support only 28 species of caterpillars (divide 20 into 557). This may be theoretically possible, but one would think that, by now, a great deal of host specificity would have been noticed in oaks.

Supporting the broad use of oaks by caterpillars is the leaf damage cater-pillars leave behind. Two summers ago, one of my students, Christian Stoltz, compared total leaf area consumed on 16 different species of oaks in south-east Pennsylvania. He found no significant differences in the amount eaten among fourteen oak species. Two species, willow oak and water oak, had slightly less damage than the other species, but these two species are planted north of their natural range and therefore beyond the reach of caterpillars that would normally eat them in the south.

I have made my arguments using oaks as an example, but these argu-ments apply to all of the plants in an area. The number of caterpillar spe-cies recorded on a plant genus is the best indicator we have of the ecological value of that plant to local food webs. One might worry that this host index will undervalue good native plants, but before we had this index, we were overvaluing many natives in terms of their contribution to food webs. The reality is that all natives do not contribute equally. Therefore, it is important to include the best contributors, not exclusively, but as the foundation plantings of our landscapes.

In your books and talks, you always focus on moths, but aren't butter-flies important too?

Butterflies are beautiful, but they cannot hold a candle to moths in terms of ecological contributions, especially to food webs. We used to think that but-terflies were a monophyletic group originating from a common ancestor only distantly related to moths, but DNA analyses have shown that butterflies are simply bad-tasting, day-flying moths. They evolved from moth lineages well after most moth lineages had diversified. They can fly during the day without fear of being eaten by birds because most are slightly or greatly distasteful, just like the iconic monarch. Moreover, there aren't very many butterflies compared to moths: for every species of butterfly in North America, there are 19 species of moths. Most people labor under the misconception that butterflies are important pollinators. A few species are, but most species simply visit flowers for nectar without transferring pollen from the male flower parts to the female structures. We love butterflies primarily because most are beautiful.

In contrast, moths have been described by eminent entomologist Dan Janzen as the "meat and potatoes" of terrestrial food webs. What he means by this is that moths, mostly in their larval stage (caterpillars), transfer more energy from plants to other animals than any other type of plant eater. Remember that the energy that drives nearly all life on Earth comes from the sun. We can't eat sunlight, though, so that energy is useless to us until it is locked up in the carbon bonds of simple sugars and carbohydrates that are created via photosynthesis in plants. In other words, the fact that plants turn sunlight into food enables the vast diversity of animal life that has evolved on this planet to exist—that is, those animals exist only if they can eat plants or eat something that ate plants. Most vertebrates do not get their energy directly from plants; they get it by eating insects that have eaten plants and have conveniently turned plant parts into valuable protein and fats. And moth caterpillars do this in more ways and at a faster rate than other types of insects. In short, I focus on moths and their conservation rather than butterflies because without moths, life on our terrestrial planet would be a mere shadow of its current self.

What size area is required to sustain a population of a species?

In the language of conservation biologists, you are asking what it takes to sustain a "minimum viable population." It takes an area large enough to contain all of the resources required to sustain enough individuals to reproduce as fast as individuals are dying—that is, enough clean water, food, and safe sites. For a territorial species, the area must be large enough to contain the territories of all of the individuals in the population. How big this area needs to be will depend on the species in question. A medium-sized compost heap or rotting log is large enough to sustain a population of pill bugs. But it takes a mature forest half the size of Costa Rica to sustain a viable population of Harpy Eagles and even more space to sustain jaguars. Many butterfly populations can be sustained within an acre or two if that space is well-endowed with host plants and nectar sources. Most of the biodiversity that runs the ecosystems that sustain us requires more space than the average lot size, which is why it is so important for everybody to accept their inherent responsibility to care for our ecosystems. Although one well-landscaped lot can help migrating birds as they pass through, entire neighborhoods that favor productive native plants are needed to sustain a population of breeding birds and passersby.

In your book *Bringing Nature Home*, you state that up to 90 percent of all plant-eating insects are considered specialists. So is it correct to assume that some of the most successful non-native invasive insects are generalists? I'm thinking of Japanese beetle, gypsy moth, and spotted lanternfly, for example. Furthermore, do these successful non-native insects prefer plants from their native range? I have heard anecdotal accounts that spotted lanternflies, for example, are especially attracted to species of *Ailanthus*.

The case of introduced insects can be complex. Japanese beetle is a specialist on grass roots as a larva but a generalist as an adult. And we have lots of Japanese beetles because in the lower 48 states, 44 million acres of lawns serve up the larvae's favorite food: grass roots. Hemlock woolly adelgids (on hemlock), emerald ash borers (on ash), and spongy moths (once known as gypsy moths, on oak) are all specialists. In specializing on these plants in their native lands, these insects acquired the ability to circumvent host defenses. That's what specialists do. But because North American species of hemlocks, ashes, and oaks have never encountered these pests in evolutionary time, these trees have not developed defenses that would deter these insects (think of the human epidemics of measles in the Hawaiian Islands and smallpox in North America). Spotted lanternfly prefers species of *Ailanthus*, but because this sucking insect can avoid many plant defenses by plugging into phloem, which is poorly defended, it does well on plants with poor phloem defenses—and there are many of those. Back to spongy moths. In specializing on oaks, spongy moths have developed the ability to circumvent tannin defenses. Several tree species use tannins as defenses, so when spongy moths run out of oak foliage as they become large larvae, they can move onto other plant species and eat them as well.

What are the seasonal variations in bird use (for food) of oaks and other trees and native plants? Insects are important, but so are berries and seeds—but at different times and to different degrees for different birds.

Most birds can be divided roughly into four diet guilds: frugivores (birds that derive most of their carbohydrates, fats, and proteins from fruit), granivores (strictly seed eaters), insectivores (strictly arthropod eaters), and omnivorous birds that eat insects and spiders whenever they can (but particularly when they are breeding) as well as seeds and berries when they are available (primarily when insects are scarce during fall and winter). There are no true frugivores in the United States, although Cedar Waxwings eat more berries than most other species. Omnivores include many of our most common overwintering species including chickadees, titmice, cardinals, some woodpeckers, and robins. Others, like many warblers, gnatcatchers, Bushtits, kinglets, Verdins, swallows, swifts, and nightjars, are entirely insectivorous. Only 4 percent of North American bird species are true granivores, able to exist and reproduce on seeds alone. These include finches, House Sparrows (which are finches, not sparrows), doves, pigeons, and crossbills.

Granivores are generalists and will consume most types of seeds they encounter, whether they come from a native plant or not. For that matter, insectivores will take most insects, but they clearly favor caterpillars when migrating and feeding young. In fact, 96 percent of North American terrestrial birds rear their young on insects and spiders. My former student Ashley Kennedy found that in 16 of the 20 most common bird families, caterpillars dominate the nestling diet. And that is the main reason native plants, particularly our keystone plants, are so important for breeding birds: they support a lot of caterpillars.

Nearly all temperate zone birds breed in spring or early summer and rely heavily on the plants that produce those caterpillars, as thousands of caterpillars are needed for each nest. Migrants also depend on copious supplies of caterpillars to fuel their arduous flights. This is especially true for spring migrants, which enter lands in which plants have yet to make berries and orthopterans such as grasshoppers and katydids are still in the egg stage. Fall migrants continue to use the fat and protein supplied by insects, but many species will also gorge on ripe, high-fat berries if they can. Most birds that overwinter either

A Golden-crowned Kinglet searches winter branches for caterpillars.

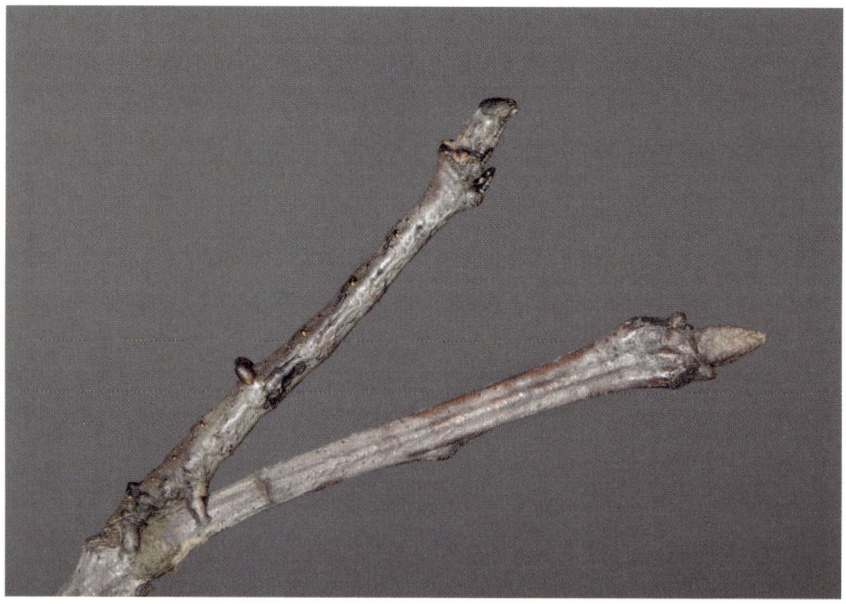

Many caterpillars, particularly inchworms, spend the winter motionless in trees.

eat seeds exclusively (granivores) or add seeds to their insect-based diets. For example, only 50 percent of the winter diet of the chickadees and titmice at your feeder is seed. The other 50 percent comprises insects and spiders, even in winter. And kinglets, surprisingly, do not migrate to the tropics where insects are plentiful all winter long. Instead, they stay up north hunting caterpillars and other insects that remain in leafless trees throughout the winter months. Insects are an important food source for most birds year round, while berries become important for some birds in summer, fall, and winter.

Do you know, or could you guesstimate, how long it would take an insect to coevolve with a host plant? I recently read that the West Virginia white (*Pieris virginiensis*) is laying its eggs on garlic mustard, which is toxic to its offspring. The article blamed this, in addition to other reasons, for the decline of the butterfly. Since garlic mustard is so persistent, I wonder if the West Virginia white can coevolve?

Garlic mustard is related to the West Virginia white's native host plant, toothwort, in the mustard family. That's why the butterfly is mistakenly laying eggs on it. How fast it responds to strong selection pressure like this will depend on how much genetic variation exists in its gastric physiology. If there isn't much variation, it will have to wait for mutations to occur. That's a very slow process. If there is a lot of variation, it can evolve fairly quickly, in maybe 20 generations.

I have heard you mention winter moths, a group of moths that fly during the winter months. How can they be active when it's cold?

Several species of moths, including the sloping sallow, bicolored sallow, and roadside sallow, don't emerge as adults until mid-autumn, and they are active right through Christmas. These moths are credited with being the primary pollinators of witch-hazel, a plant that does not bloom until after the leaves have fallen in a deciduous forest. Other species, such as the joker moth, Comstock's sallow, and major sallow, emerge in late winter and are active until mid-April. And yet another group of moths are active from October all the way until May. These include many species of *Lithophane* pinions, mustard sallow, and straight-toothed sallow.

The general lore is that winter moths will fly on any night with a temperature above 50 degrees Fahrenheit, but I have collected them at my house when it was 45 degrees and falling, and I've watched a figure-eight sallow (*Psaphida resumens*) fly into a light when it was 32 degrees! How can a cold-blooded moth be active at such low temperatures? By shivering! Really! They vibrate their thoracic flight muscles rapidly enough to generate heat. This self-generated heat enables them to stay active when other insects are in diapause or deep torpor. Winter moths occupy a nearly empty niche by flying during the cold months, but it is hard to imagine why natural selection favored such behavior in so many unrelated species. Maybe being active after most of the insectivorous birds have migrated south was a good enough reason, but it seems like a hard life to me. (I get cold easily!)

Winter moths such as the bicolored sallow are credited with pollinating witch-hazel.

Are riparian corridors good for migrating birds?

Of course! Riparian corridors (well-planted buffers along waterways) are beneficial for lots of reasons, including habitat connectivity, shading the waterway to reduce the water temperature (many fish and mollusks cannot tolerate warm water), and filtering pollutants before they enter our streams and rivers. But they can also supply habitat (a place to live and food to eat). Some rivers—most notably the Mississippi—flow from north to south, the same direction migrating birds are traveling. These rivers provide road maps for migrants, and habitat along the waterway provides essential stopover resources. When flying over land, most migrants travel only at night, and they don't fly more than 300 miles in a single night. After every flight, they must stop and rest wherever they are. Whether or not the riparian corridor lies in the direction they are flying, it can serve as a stopover site full of insect food in spring and insects and native berries in fall if it contains the appropriate native plants.

I have always brushed off spider webs from my outdoor structures whenever I see them. I've heard that's not a good idea.

Right. Spider webs are important resources in nest construction for small birds such as hummingbirds, Blue-gray Gnatcatchers, kinglets, and Bush-tits. Spider web silk is made of complex proteins that are elastic and tough. When spider silk is a major component of a nest, the nest will stretch to accommodate the size of the nestlings within it as they grow. Spider webs are also the homes of the spiders we want, if for no other reason than to eat the mosquitoes we don't want.

Since adult moths are an important food source for bats, do we have any idea how bird predation of caterpillars affects the feeding and breeding success of our bat populations?

To my knowledge, no one has attempted to demonstrate experimentally the connection between moths, birds, and bats, though logically we know it must exist. If A eats B, and C eats B, then there is the potential for A and C to compete with each other for access to B. It's not just birds and bats that depend on moths, as countless other predators and parasitoids do as well.

This is just another reason to plant species that support lots of Lepidoptera to ensure that there are enough caterpillars and moths to meet the needs of organisms at higher trophic levels.

Which caterpillars are birds focusing on early in the season to feed their young? I don't see any caterpillars out there, and I wonder what species are active early on?

Migrating birds focus primarily on caterpillars in the Geometridae family, the inchworms. Not only do they hatch out just as leaves are expanding, but many species overwinter as nearly mature caterpillars and can be found before any leaves pop. You don't see any spring caterpillars for two reasons: the birds have eaten most of them and there are fewer caterpillars these days than when I was growing up.

I am worried that if I allow insects to eat my plants as you suggest, they will kill most of them. Do I worry too much?

You certainly do! Who do you think was preventing insects from killing all the plants before we arrived on the scene? From all I've heard, North America was a vibrant, verdant place even before humans first set foot on the continent, and with no help from us. That's not to say that nothing was controlling phytophagous (plant-eating) insects. In all of her wisdom, Mother Nature did not create ecosystems with just two trophic levels—plants and the creatures that eat plants. She also included predators, parasitoids, and diseases that live off insect herbivores—thousands of species that keep insect herbivore populations in check.

The ironic thing about spraying to keep your plants safe from insects is that the first creatures to die from insecticides are the insect predators and parasitoids that control plant eaters. Because they do not have to detoxify defensive compounds in plants, as insect herbivores have to do every day, insect natural enemies are far more sensitive to insecticides than are the target insects. Often, spraying actually creates more insect problems than it solves. Spraying favors pests that are resistant to insecticides, while killing the natural enemies that would have controlled potential pests. Your plants are far more resilient to insect herbivory than most people think. The best course of action when you have an outbreak is to do nothing and let the natural enemy community do its job. If you want to worry about something worthwhile, try worrying about the health effects of the insecticides you use to kill those insects that are eating your plants.

Does a higher diversity of caterpillar species hosted by any one tree genus always translate into a greater abundance of caterpillars in any one tree during a given year?

A strong correlation exists between species diversity and abundance. For example, oaks support the most species of caterpillars in North America and, on average, they also support the greatest number of caterpillars in a given place at a given time. There are important exceptions to this general rule, and the spruce budworm in Canada's boreal forests is probably the most blatant. The spruce budworm is a complex of species in the genus *Choristoneura* that develop on conifers in the northern part of the continent. In any one area, usually only one or two species are present, but they build to enormous numbers so reliably that many warbler species choose the boreal forest as their breeding destination. Species such as the Bay-breasted Warbler assess the size of the budworm population and then adjust the number of eggs they will lay in their nests accordingly. When budworm populations are high, these warblers can successfully raise up to nine chicks in a single breeding attempt. In contrast, when budworm numbers are low, they may lay only four eggs and then struggle to feed their chicks.

How do moths find their host plants? It seems inconceivable that they would be able to find a new planting somewhere.

It does seem incredible that moths can locate their host plants within a matrix of all the other plant species growing in a particular area, which can easily number in the hundreds. But it is no more incredible than it would be for you to locate a person in a crowded room who is wearing your favorite perfume or cologne; you would simply walk around and sniff each person you encountered until you found that fragrance you love. Moths and other insects that eat plants find the appropriate plant species in the same way— by smell. And they are very good at it, because the very existence of their species has depended upon their ability to detect and recognize the odors of their host plant accurately for millions of years.

Just like our noses, insect bodies are loaded with thousands of chemoreceptors. Insect antennae are packed with receptors capable of detecting minute quantities of odors produced by plants or, if they are males, by females of their species. A female moth looking for the right plant on which to lay her eggs, for instance, will follow odor plumes of her host plant as she flies through the air, until she finds a likely candidate. She must be certain she has chosen correctly, though, for if she lays her eggs on the wrong plant species, her young caterpillars will not be able to eat that plant and they will all die. So she employs a second set of chemoreceptors on her feet. She lands on the plant and smells, if you will, the signature odor of that plant, which is embedded in the waxy coating covering its leaves, by walking over the leaf surfaces. Only when she is satisfied that she has indeed arrived at an acceptable host plant will she lay one, several, or all of her eggs.

How many leaves of a plant are required to "make" a butterfly or moth? How many caterpillars does it take to "make" a bird?

This is a wonderful question, because it forces us to think about the mechanics of a viable food web. Having food available for members of each trophic level is a good start, but it results in a successful food web (the transfer of energy from plants through all the layers of herbivores and predators) only if there is *enough* food available to support vibrant populations of all groups. So how much is enough? How many leaves of a plant are required to bring a caterpillar through from egg hatch to pupation? Someone actually calculated that number for spicebush swallowtails, and it turned out to be three mature spicebush leaves—that is all it takes to make one spicebush swallowtail. You could have a large population of spicebush swallowtails in your neighborhood and never notice any of the herbivory required to make that population on your spicebushes. Some caterpillars are messy eaters and accidentally snip off a few sections of the leaf they are eating without actually eating those sections. Those species would require a few more leaves to reach maturity. And these estimates would depend on how big the caterpillar gets before it pupates. Cecropia moth caterpillars, for example, reach the size of a sausage! Naturally they eat many more leaves than a small moth like the distinct sparganothis, a

Cecropia moth caterpillars can easily be 5 inches in length when fully grown.

specialist leafroller on goldenrod. In all cases, however, a single caterpillar—even one of our largest caterpillars—requires only a small portion of its host plant to complete its development.

How many caterpillars are required to make a bird? Few people have actually sat down to count all of the caterpillars that birds bring to their nest, but those who have had the patience to do this have recorded astounding figures. In 1973, for example, Robert Stewart made detailed records of a Wilson's Warbler pair while they were feeding their young. He found substantial differences in how diligently the male and female worked at this endeavor. The male was no slouch, carrying food to the nest 241 times in a single day, but the female put him to shame: on that same day, she fed the nestlings 571 times! This rate was maintained over the five days he watched the nest. Stewart did not count the actual number of caterpillars the pair brought to the nest; feeding was rapid and often a parent carried more than one caterpillar in its beak at a time. Yet even if only one caterpillar was brought to the nest each trip, the pair would have brought in 812 caterpillars per day, or 4060 caterpillars in total, during the five days Stewart watched the nest. The chicks he observed stayed in the nest for eight days before they fledged.

These observations are not exceptional. In 1967, Louise de Kiriline Lawrence recorded sapsuckers feeding their young 4260 times, Downy Woodpeckers 4095 times, and Hairy Woodpeckers 2325 times before the young fledged and left the nests. In 1971, Stephen Martin watched Bobolinks bring food to their nests 840 times a day for 10 days in a row. All of these species regularly bring in multiple prey items per trip. The point is clear: to have breeding birds in our ecosystems, we need to grow plants that generate huge numbers of caterpillars.

CHAPTER TWO

Biodiversity

**Can we expect what's happening on Guam, with the invasive
brown tree snake species destroying local ecosystems, to occur in
mainland ecosystems?**

Island ecosystems are more fragile than mainland systems because they
always contain far fewer species. The loss of one species often triggers the
loss of several others. There are significant losses of biodiversity on Guam
because of the introduction of the invasive brown tree snake, which preys on
and has eliminated nine of the eleven native forest bird species on the island.
Because many of the island's tree species depend on those birds to spread
their seeds, the forests are suffering as well. Yet you see human life continuing
unhampered. Why is that? Because humans on Guam have never depended
solely on the island for food, fiber, wood products, and other essential natural resources.

If ecosystem collapse on Guam is foreshadowing the collapse of mainland ecosystems across the world, we had better take note. Humans will suffer, because there will be no influx of vital resources from some other planet,
the way resources now flow to Guam from mainland ecosystems. Eliminate
pollinators, such as much of China has done, and you will see a greater and
unhealthier reliance on wind-pollinated crops like wheat and corn. The collapse of forest ecosystems across the world would negatively impact climate
and carbon cycling. Lose native plants, and the insects that fuel higher trophic
levels will also be lost, with the subsequent loss of most other animals. Maybe
we humans can survive for a while in a world with only wind-pollinated
plants and a few insects to recycle nutrients, but such a world will be unstable,

and the planet's carrying capacity, already drastically lowered by humans, will fall even more. It will create a downward spiral that won't come close to supporting the eight billion people who already exist and certainly not the projected ten billion in the future. If that's the future we insist on, then we deserve everything we're going to get. Too bad we'll take most everything else with us. But we do have a choice in this matter! We can accept the demise of biodiversity as inevitable, or we can fight to preserve it. I know where I stand.

What can we plant that will help nighthawks and swifts?

We should plant to produce the food that these birds eat. But nighthawks and swifts have very different diets. Nighthawks depend heavily on nocturnal moths—the bigger the better. So establishing plants that produce moths, particularly giant silk moths, large geometrids, and sphinx moths, is the best thing you can do. Over much of the United States, that would mean planting oaks, native cherries, hickories, poplars, and cottonwoods. And we must either turn off our outdoor lights at night or replace the white bulbs with yellow bulbs that do not attract large moths. There is no point planting moth-producing plants if night lighting lures moths to their death.

Swifts, in contrast, are aerial gleaners, taking just about any insect they can catch while flying. Although they catch high-flying leafhoppers and ballooning spiders that are in the air during the day, they depend more on nematoceran flies such as midges, crane flies, and mosquitoes that are produced by the millions in healthy wetlands. Keeping our rivers, streams, and marshes clean and productive is the best thing we can do to provide food for swifts.

What can I do to attract bats?

To attract bats, provide lots of bat food. In the United States, that means nocturnal insects. Along with ending our age-old war on insects, protecting existing wetlands or building new ones is a great way to encourage populations of the chironomid and chaoborid midges, caddisflies, and mayflies that bats eat all season long. We must stop killing tent caterpillars and fall web worms that turn into the night-flying moths that nourish bats. Lobby to end mosquito fogging in your township—it kills all kinds of insects but does not control mosquitoes.

A necessary part of this nighthawk's diet is large moths.

Unfortunately, increasing populations of the insects that bats require is not guaranteed to bring bats to your yard immediately. That's because white-nose syndrome is decimating bat populations. Since the winter of 2007–2008, about 90 percent of what used to be three common bat species have died in 38 states and eight Canadian provinces from this infectious disease. The fungus thrives in the cold and humid conditions of caves and mines used by bats. Brought to North America from Europe and/or Asia by spelunkers, *Pseudogymnoascus destructans* infects skin of the muzzles, ears, and wings of hibernating bats. The good news is that some populations, particularly little brown bats, seem to be adapting to the fungus and their numbers are slowly increasing.

What is the value of vegetable gardens with respect to biodiversity and ecosystem function?

Vegetable gardens do not contribute directly to biodiversity or ecosystem function in any meaningful way, but that is not their purpose. We grow vegetables at home for fun, to provide fresh veggies with a known pesticide history, and to reduce food transportation costs. Growing vegetables that have not been flown or trucked in to your local grocery store from faraway states or countries keeps carbon out of the air. In that sense, your vegetable garden does help biodiversity by helping to reduce climate change impacts. That may seem like a stretch, but every little bit helps.

Insect Declines

It is well-publicized that monarchs and honey bees are in trouble. Are any other insects in trouble?

Yes, the North American monarch, now red-listed by the International Union for Conservation of Nature, has lost 99.9 percent of its population in the West and about 85 percent of its population in the East as of 2021. And colony collapse disorder is a serious threat to honey bees, which have taken a beating across the country since in the early 2000s. But the new era of concern for insects in general began when a 2017 study by Caspar Hallmann and colleagues reported declines exceeding 75 percent in the biomass of flying insects across Germany over the course of three decades. This study stimulated a flurry of worldwide research, and dramatic declines in insect abundance and diversity have now been reported from many countries. A decade ago, a study by Rodolfo Dirzo and others estimated that we had already lost more than 45 percent of our insects! The results of these studies are no surprise to old-timers like me, who can clearly remember insects splattered on car windshields and grills, the blizzard of moths in our headlights as we drove along country roads at night, and the frenzied mass of insects twirling around streetlights during summer. But those days are gone.

Decades ago, E. O. Wilson made general but dire predictions about the ecological consequences of insect declines, including the loss of flowering plants, terrestrial food web collapse, and the associated loss of animal diversity. His message was clear: insects are essential components of most terrestrial ecosystems that are critical for the maintenance of those ecosystems as well as populations of birds, reptiles, amphibians, and mammals (including humans). Translation: Insect declines are not something we can live with for long. They must be reversed.

The causes of insect declines are many—so many that David Wagner of the University of Connecticut likened insect declines to "death by a thousand cuts." Industrial agriculture, insecticide misuse and overuse, non-native plants, light pollution, mosquito zapping and fogging, climate change, and of course habitat loss are all taking their toll. To me, there is hope hiding within this list, however; with the exception of industrial agriculture and climate change, we as individuals can effectively address each of these causes right where we live. We can throw away our needless insecticides; we can insist on vegetable seeds that are not coated with neonicotinoids; and we can fire the mosquito fogging company that, despite its claims, does not control mosquitoes but instead kills all of the other insects the fog comes in contact with, including monarchs and pollinators. We can throw away our bug zappers that kill thousands of insects nightly—only 0.02 percent of which are biting flies. We can curb our addiction to outside night lights, either by turning them off entirely or by replacing the white bulbs with yellow ones that do not attract nocturnal insects. And we can fight habitat loss by reducing the area of our properties wasted on lawn and replacing non-native ornamentals that do not support insect development with the native plants that do. Insects are impressively resilient. Give them what they need to reproduce and they will rebuild their populations faster than you might imagine. We *can* reverse insect declines if we make it a priority.

Are insect populations really declining? I've seen scientific papers questioning the prevailing view.

Yes, I have seen those papers too. They are reporting that in several monitoring stations around the world, insect populations are stable. This is good news, but it misses the point. Insects are typically monitored within protected, undisturbed sites over long periods. And, indeed, when such sites are considered alone, studies show no strong patterns of decline. Unfortunately, most of the globe is not protected and undisturbed. As David Wagner (in a 2021 study) says, you cannot add eight billion people to a finite planet without causing insect declines. Globally speaking, less than 25 percent of the land has not been directly modified by human activity. Statistics compiled by the United Nations' Food and Agriculture Organization tell us that each year, 10 million hectares (24.7 million acres) of forests are cut and converted into agriculture, housing, and other forms of human "development." There were 1.2 million housing starts in the United States alone in 2021. I could go on ad nauseam with such depressing statistics, but suffice to say, such habitat destruction has been the primary driver pushing species into steep declines.

Matt Forister (University of Nevada) and his colleagues offer a useful analogy in a study published in 2023. Natural locations, where most insect monitoring occurs, are like life rafts leaving a sinking ship. Skeptics of insect declines are making a mistake by equating what's happening on the life rafts with what's happening on the ship. To measure insect declines accurately, we have to monitor the sinking ship as well. And to make matters worse, climate change is threatening even the life rafts. Drought and heat in western North America, as well as floods and torrential rains in the east, are taking their toll on insect populations, even in protected areas. So, yes, insect populations, along with most other forms of essential biodiversity, are declining.

When I was young, grasshoppers and crickets were everywhere. Now I see very few. Do you know why?

Whether or not grasshoppers, crickets, and other orthopterans are disappearing depends on where you are. Fortunately, grasshoppers and crickets are doing well in many places, but, as you note, they are nearly gone in many others. Anyone who uses a lawn-treatment company to kill insects and fertilize lawns will not have crickets or grasshoppers in their yard. Anyone who uses herbicide "turf builders" and mows religiously will have very few grasshoppers and crickets in their yard. Any farmer growing roundup-ready corn or soybeans will not have the plants needed to support grasshoppers and crickets on their land. And we are just now learning about the negative effects that artificial lighting at night has on local insects.

Millions of acres that are now lawn in the United States once supported the native herbaceous plants that fed lots of grasshoppers and crickets.

Grasshoppers, despite their name, depend primarily on broadleaf forbs, while crickets develop mostly on dead plant material. In pursuit of our obsession for neat landscapes, we have eliminated both insects in way too many places. Finally, areas overrun with invasive groundcovers such as Japanese stiltgrass, vinca, or English ivy won't support grasshoppers because the plants grasshoppers depend on have been replaced by species they cannot eat. We can restore populations of grasshoppers and other insects if we plant more of our private and public spaces with the native grasses and forbs they depend on.

Do we need to create water features to help terrestrial insects?

Water features aren't needed in the eastern parts of North America, where rainfall is usually plentiful. Insects do drink water, but morning dew is enough to meet their needs all day long in these areas. In the arid West, however, yard water features are a welcomed resource for birds and insects alike. It is common to see honey bees lined up along a water source having a good drink during dry spells.

Is climate change decoupling insects from their host plants?

You would think that it might, but so far there is little evidence that insect herbivores have been unable to track developmental changes in their host plants. Unlike migratory birds that time their migration primarily by changes in day length, the timing of insect life histories is governed by the same environmental cues that control the timing of plant development. So if warmer winter temperatures trigger early leaf expansion in a plant, they will also trigger early emergence of the insects that depend on that plant. Many insects track host plant phenology directly by inserting eggs into the vascular system of a plant. This is the case for most treehoppers, for example. Egg hatch, then, is triggered by sap flow, directly linking the insect's life history to seasonal changes in their host plants. Climate change *is* impacting insect herbivores in other ways, though. When droughts and floods kill host plants, the insects tied to those plants suffer and populations decline.

You claim that the use of non-native ornamentals and invasive plants is a cause of the global insect declines many scientists are measuring, but do you have any direct evidence?

I have no direct evidence but plenty of indirect evidence. Directly testing the hypothesis that non-natives have contributed to insect declines would require a controlled, globally replicated experiment lasting at least 40 years (because insect populations vary widely, you'd need long-term datasets to show trends), in which one treatment includes only native plants and the other treatment (or treatments) includes varying degrees of non-natives. Treated areas would have to be large, because insects move around quite a bit and the amount of plant biomass would have to be the same in both treatments. Naturally, this hasn't been done and probably never will be, because it would be nearly impossible to control for these conditions. But I have used simple logic to draw the conclusion: if A = B and B = C, logic tells us that A = C. My students and I have shown time and again that insects do poorly on non-natives and local populations are reduced by as much as 96 percent when non-natives replace natives (A = B); it is also well-documented that non-natives have replaced natives over many, many millions of acres globally (B = C). Therefore, it is not unreasonable to conclude that non-natives have depressed insect populations over millions of acres (A = C).

Many people accept the evidence that non-natives do not support insect herbivores well at all, but what most people don't realize is the extent to which non-native plants have replaced the natives that do support insects around the world. My colleagues and I reviewed this subject in 2020 with lots of evidence and citations. But the extent of the problem is so large that I will give you a teaser here.

Non-native plants are everywhere! That may seem like an overstatement, but it is all too accurate. Although humans have always carried plants with them as they moved about, most introductions have happened in the last 200 years. And the rate and scale of these introductions is staggering. An estimated 13,168 plant species (about 3.9 percent of global vascular flora) have been introduced and naturalized beyond their native ranges as a result of human activity, with more than 3300 species of non-native plants established in the continental United States alone.

Because of the popularity of non-native plants in landscaping, horticulture is a major source of non-native plants in both cultivated and natural ecosystems. There are estimates that 50 to 70 percent of invasive and naturalized species are a direct result of intentional horticultural introductions. Even though many invasive plant species are regulated in the United States and included on "do-not-plant lists," these problematic taxa are abundantly sold in horticulture; in fact, 61 percent of non-native plants on the invasive species list are still sold in all lower 48 states.

In horticulture and ecological circles alike, concern has focused primarily on invasive species, with the assumption that if a plant is not invasive, it does not cause ecological problems. Indeed, the majority of ornamental species have not become invasive, leading land managers and the public to deem these species acceptable for planting. There are two problems with this reasoning, however. First, ornamental plants that are distributed by the millions across landscapes represent the plant base wherever they are used. Even if they never develop invasive behavior, they have not replaced the ecological functions of the native plants that used to support insect populations. Second, there is no guarantee that an ornamental species that is well-behaved today will not become invasive in the future. Many invasive plants experienced a lag phase during which they were benign or overlooked members of plant communities before being recognized as invasive.

The need for fast-growing colonizing tree genera such as *Acacia* and *Albizia* in agroforestry and restoration has also increased the spread of non-native species, many of which have escaped to dominate nearby native forests. At least 25 percent of the world's planted forests comprise non-native tree species. One-fourth of Portugal's forestland (900,000 hectares) is planted in non-native eucalyptus. At least 118 exotic tree species have naturalized in Puerto Rico and compete with native species in natural stands; the African rubber tree, a species that is invasive in more than 30 countries and completely dominates secondary forests of many tropical Pacific islands, is now one of the most common trees in Puerto Rico as well.

**Do you really believe that non-native plants could possibly threaten
the continued viability of necessary insect populations? It really seems
far-fetched to me that there could be enough of these and other plants to
threaten your insects.**

First, let me emphasize that science is not a discipline based on beliefs, but
one based on testable hypotheses. I have tested many times the hypothesis
that non-native plants support fewer insects (both insect herbivores and pollinators) than do native plants. I always get the same answer: they do. And
because so many studies have shown the same thing, favoring native plants
over non-natives is hardly far-fetched. There has been a 45 percent decrease in
global insect populations since the 1970s. You refer to insects as "my" insects.
They *are* my insects, but they are *your* insects as well. Regardless of whether
you like or appreciate what insects do for you every day, you would not be
here without them. E. O. Wilson has called insects "the little things that run
the world," and without them, we would lose 90 percent of our flowering
plants, which would cause the collapse of nature's food webs and the loss of
birds, mammals, reptiles, and amphibians. The loss of insect decomposers
would end the rapid recycling of nutrients, leaving only fungi and bacteria to
break down organic matter. We humans would not last long in a world with
no insects.

The good news is that there *is* room for compromise in our plant choices,
and I am hoping we all can embrace such compromise to create ecologically
productive landscapes without destroying the nature we all need or the joy
that gardening with exotics brings to so many people. My lab has found
that when 70 percent of the plant biomass in a landscape is native, enough
insects are present so that birds can successfully breed in our yards. We also
know that woody plants support far more insects than do herbaceous plants.
Finally, we now know that just 14 percent of our native plant species support
90 percent of the caterpillars that drive local food webs. We call these powerhouse plants "keystone species" because they are essential components of
every landscape. That said, if keystone species are present, a few lilacs, crape
myrtles, camellias, decorative annuals, and the like do little harm.

CHAPTER THREE

Native and Non-native Plants

What's the one best statement to convince others to plant natives?

Native plants support the life that runs the ecosystems that support us. Plants from other continents do not. An exception here and there does not change the generality of this statement.

I am an avid gardener but often find that the expense of purchasing native plants is out of my budget. Do you have any suggestions?

I often use plantings on my own property to illustrate how easy it is to "go native" on a low budget. Cindy and I own 10 acres, and we have restored them with more than 120 species of woody plants and who knows how many species of herbaceous plants for less than 200 dollars. I planted some species (mostly our oaks and beeches) from seed I collected, but the rest came here through what garden designer and author Rick Darke calls "addition by subtraction." When a plant came in naturally, carried and distributed by Blue Jays, squirrels, deer, wind, or the existing seed bank, we left those that we wanted to grow. And we removed those that we didn't want, primarily all of the invasive species that were on the property when we bought it. I've purchased a few native species at plant sales associated with talks I have given, but this wasn't necessary and I did it mostly to support local growers. Your challenge is keeping the bad guys out, not attracting the good guys. They will come free of charge!

Are walnuts good at supporting biodiversity?

Black walnuts are not the highest ranking plants, but they are nevertheless solid contributors to local biodiversity. Each year they produce nutritious nuts that sustain squirrels and other rodents throughout winter. The same goes for related butternuts. In most eastern states, walnuts and butternuts serve as host plants for more than 100 species of caterpillars, which makes these trees good at providing essential food for baby birds. Walnuts are also good at creating a dark brown, beautifully grained wood, prized by woodworkers everywhere. But walnuts and butternuts do have a downside: both produce juglone, a compound that is at least mildly toxic to many vegetable crops, perennials such as crocuses and columbines, and valued ericaceous plants such as azaleas, rhododendrons, and blueberries. Luckily, dozens of plants are not affected by juglone, including viburnums; smooth, winged, and staghorn sumac; ninebark; wild roses; blackberries; hazelnuts; dogwoods; oaks; Virginia bluebells; and fringe trees. So whether a black walnut or butternut

Many plants, like these Virginia bluebells, can grow well under black walnut trees.

is the right tree for a particular space depends on several factors—it's a judgement call for the land manager. If you want to grow trees quickly and easily on your property, walnuts are for you. And squirrels spread them around. My problem at home is keeping the walnuts at bay.

Do boxelders support as many caterpillars as maples?

Boxelders are distant relatives of maples but do not support caterpillars as well as most other maple species. Many people consider them weedy because they tend to pop up where you don't want them. But they do well along streams and in wetlands (as a riparian species) and can grow into large, beautiful, spreading trees if given space and sun. If I had limited space, boxelders would not be high on my list to plant; otherwise, I would welcome them as part of the native diversity in most North American ecosystems.

Do I need to improve the soil before planting natives?

Probably not. No one was improving the soil for natives before we got here, and they all did just fine. Gardeners are often tempted to change soil qualities by adding compost, sand, or fertilizer to force a particular plant into a spot where it doesn't naturally belong. One of the beauties of native plants is that you don't have to do that. You *do* have to choose plants that naturally thrive in the type of soil you have, however. If you have acidic soil, plant acid-loving natives such as azaleas and many oaks. Likewise, if your soil has a high pH, focus on species such as California lilacs (*Ceanothus* spp.) and native viburnums that do well in basic soils. Right plant, right place applies to native plants as much as it does non-natives. Finally and whenever possible, soils that have been compacted by farm machinery, mowers, trucks, and the like should be broken up mechanically with a ripper, pickaxe, or fork before planting.

I've been told by fellow master gardeners that annuals are just weeds and have no role in serious restorations. Is there any truth to that? My instincts say otherwise.

Your instincts are entirely correct. Native annuals are every bit as important as native perennials. A weed is a plant out of place, and native annuals are not out of place. *Bidens aristosa* (ditch daisy), for example, is a beautiful yellow-flowered annual "weed." The goldenrod stowaway moth, as a specialist, can't live without it and would be very sad if this annual were eliminated from our landscapes. Fifty-five species of caterpillars in my county use species of *Ambrosia* (ragweed), and this early successional plant is host to many of them, such as the beautiful common spragueia. Ruby-throated Hummingbirds would not agree that common jewelweed is worthless, and annual horseweed supports the brown hooded owlet, an extraordinary caterpillar. There are countless other examples of how annuals support valuable parts of our biodiversity, but you get my point. Your years of rewilding experience have imparted great wisdom and common sense.

Is there such a thing as a "good" or "bad" native plant?

Good or bad at what? Are there native plants that will take over large areas, expelling all other species the way so many invasives do? No. But there are native plants we can call "aggressive," including Canada goldenrod and Virginia creeper, and we should plant them with that in mind. At the same time, Canada goldenrod and Virginia creeper both contribute more to ecosystem function than most other plants. Larger, longer lived plants often have bigger impacts, simply because of their size and age. Trees with large root systems manage the watershed better than tiny spring ephemerals. They also remove more carbon from the atmosphere, and, if they are a keystone species, they pass on more energy to animals. But many smaller flowering annuals and perennials support more pollinators than larger woody plants do. Different plants are good at different things. To cover all of the ecological bases, your best bet is to plant a diversity of plant types wherever you can.

How should I use Virginia creeper?

Virginia creeper is a great groundcover for shady areas. It also does well along fences or for covering a snag in full sun. If it's in the sun, it will flower and is loved by many native bees. Those flowers will then turn into berries high in fat, which are perfect for migrating and overwintering birds. It also has great fall color, so you can use it as a beautiful accent in your landscape. Finally, several species of large sphinx moth caterpillars, favorite foods for nestling

The Ruby-throated Hummingbird depends on jewelweed nectar during its fall migration.

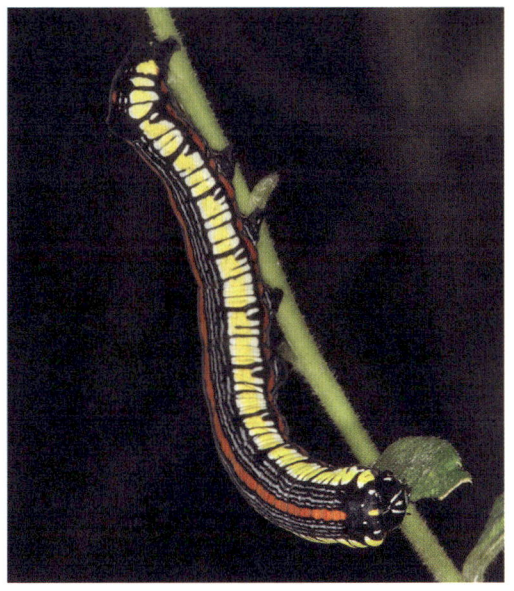

The brown hooded owlet, *Cucullia convexipennis*, often uses horsetail as a host plant.

cardinals, develop only on Virginia creeper. I would not plant it against your house, though, since the vine has aerial tendrils that can adhere to and damage siding and bricks.

Are there plants that have been domesticated for so long that they would be considered native? For example, cushaw squash was domesticated between 7000 and 3000 BC and was used by Indigenous Americans all the way up to what is now Pennsylvania and beyond, even though it originated in southern Mexico. Or, for example, native elderberries. Although there are some European varieties, the native ones are being domesticated. If I'm planting either of those, do they count as natives?

The plants that were domesticated by Indigenous people thousands of years ago were brought up from Mesoamerica slowly enough that the insects that depended on them have been able to follow. That's why there are squash bees, a specialist pollinator of cucurbits, in Delaware. There were no cucurbits in Delaware until Indigenous people brought them here, but the squash bee that so effectively pollinates them was able to track that movement. The same can be said for the squash beetle, squash bug, and melon worm. The point is this: because Native peoples planted squashes from their historical range into North American lands slowly and incrementally, much of the fauna that was already adapted to squash was able to move with it.

Two species of elderberry are native to North America: *Sambucus nigra,* which occurs pretty much throughout the country naturally, and *S. racemosa,* which has populations in the northeast and western parts of the continent but does not occur in southern or central areas. If you plant an elderberry within its historic natural range, it will contribute to your local ecosystem just as if it had been planted by a bird. The elderberries planted beyond their historic ranges will not contribute at the same level as the natives.

Why haven't the elms you planted on your property died of Dutch elm disease?

I can only guess at an answer, but here are two possibilities. The seeds came from trees on the University of Delaware campus that had not died, so maybe

The squash bee, *Xenoglossa strenua*, is a Central American species that was able to follow the northern range expansion of cultivated squash.

they bore some genetic resistance. Another possibility is that my trees are quite isolated from other American elms, so the population of blue-stain fungus that causes Dutch elm disease has been locally extirpated because of lack of hosts. In any case, the trees I planted from seed 23 years ago are now over 80 feet tall and completely healthy. And, I might add, they were free! I scooped up the seeds out of the gutter!

Do pink lady's slippers have an obligate relationship with oaks?

No. They are typically associated with conifers and are most abundant wherever pines are found naturally in the coniferous forests of New England and Canada. Doug Gill (University of Maryland) has studied pink lady's slippers for decades and has a pretty good idea about why we occasionally find these plants in hardwood forests. Those forests were fire-maintained pine savannahs a mere 60–100 years ago. Although they are hanging on to this day, pink lady's slippers will most likely disappear from hardwood forests over time.

How should I choose the right native plants for my property?

Dozens of beautifully illustrated regional guides to native plants are available for all parts of the country. And, of course, today we have the internet, with its endless supply of information about native plants. Several years ago, I worked with Mary Phillips, Naomi Edelson, and others at the National Wildlife Federation to create the Native Plant Finder specifically to address this very question. Many, many species are appropriate members of native plant communities throughout the country, but we wanted to create a tool that would enable people to identify the plants that would support the most species of caterpillars, the bread and butter of terrestrial food webs, where they live. Naomi secured US Forest Service funding to support my research assistant, Kimberley Shropshire, while she scoured the literature for host records (when they existed) for every species of caterpillar in the United States. With that information, we built a web tool that enables people to identify the plant genera in their county that contribute the most energy to local food webs, simply by entering a zip code.

I have started a series in a monthly newsletter of what can be planted instead of (name your non-native/invasive plant). One challenge is that many people still need to be convinced that non-natives contribute little to the ecosystem. I have used your numbers of species of Lepidoptera that are supported by various genera of trees (such as 557 Lepidoptera species on oaks, just in the mid-Atlantic region, and more than 950 nationwide). I'd like to contrast numbers of caterpillars supported by crape myrtles. Many people believe it is native because it is so prevalent in our county.

There are three records of caterpillars using crape myrtle! We have not tried to determine whether or not they are false records, which are more common than you might think. In any case, it's not much, and anything on crape myrtle would be rare—the plant certainly doesn't support the abundance of caterpillars required by birds. Compared to oaks in the mid-Atlantic region (557 species), black cherries (456 species), or native willows (455 species), crape myrtle is not pulling its weight. I am not a betting man, but I would be willing to offer 1000 dollars for every caterpillar you find on a crape myrtle in May or June, which is when the breeding birds need them. I am confident that I would not lose a cent!

How can we grow our own natives?

The operative word here is "grow." Transplanting natives from the wild is highly discouraged unless it is a rescue mission—that is, you're saving plants from the bulldozer. But there are many opportunities to collect seeds when they are produced. In the temperate zone, most seeds will not germinate until they experience winter conditions for at least a few months. This is easy to simulate in the refrigerator. It's called "stratifying seeds" and can be done in moist sand or peat moss, often within a zippered plastic bag. By becoming skilled at germinating your own seeds, you can save lots of money and gain access to many species that are not available in the nursery trade. And it's fun.

I'm trying to find a native crabapple tree and have not been successful. Any suggestions? Is there a cultivar that would be almost as good as a native?

Four species of crabapples are native to the United States: one in the West (Pacific crabapple) and three in the east upper Midwest (prairie crabapple, sweet crabapple, and southern crabapple). Pick one appropriate to your location and then ask your local nursery to find one for you. If they can't or won't, try another nursery. They are available in the trade and shouldn't be very hard to find. No need to choose a cultivar; they are beautiful trees as straight species.

Do you have a favorite native willow?

We have several native willow options in the Northeast. A few are small trees (black willow and pussy willow). The rest are large shrubs. These include Bebb's willow, heart-leaved willow, sandbar willow, silky willow, and prairie willow. All are great biologically because they rank just below oaks in terms of supporting caterpillars and are wonderful early-season sources of pollen and nectar for native bees. I would go for the one you can find in the trade. Black willow may be too messy for regular suburban use, but everyone loves pussy willows like *Salix discolor*.

I have a 100-year-old house with lots of non-native privet hedges.
I can't come up with a replacement to plant along 20 feet of driveway.
Any ideas?

Replacing an old privet hedge is no small undertaking. There are few options in the eastern United States, but eastern red cedar is a good one that grows densely and can be trimmed to any height. Wax myrtle (also called bayberry) can also make a great hedge.

When presented with a new idea or change, people typically wonder, how does this help me? Or how would people perceive me if I do this? What are some reasons that planting native will benefit the person doing the planting?

What you're asking is this: What are the immediate, short-term benefits of enhancing your local ecosystem with ecologically powerful native plants as opposed to degrading it with toxic lawns and Asian ornamentals that support little life? That's an important question, because we know humans typically engage in something only if they can immediately realize measurable benefits. We tend to ignore long-term benefits completely. Our refusal to take meaningful action to curb climate change is a good example. The short-term benefits of reducing carbon emissions are difficult to articulate, even though the long-term costs of not curbing emissions are deadly. Telling someone that they should plant natives to help sustain specialist bees, or to support the caterpillar populations our North American birds require to breed successfully, or to prevent the collapse of the ecosystems that supply our life support focuses only on abstract, long-term benefits.

People who have already internalized how important those long-term benefits are can get immediate satisfaction by knowing they have just contributed in real ways to the conservation effort. But those arguments often mean little to folks whose ecological IQ needs a healthy boost. Well-designed native landscapes need less maintenance once they are established, but creating those landscapes is not labor or cost free, and the initial investment of time and money to do that can make sitting on a lawn mower for a few hours every week look pretty good. This is why top-down incentives to plant natives would be a game-changer overnight. Modest tax incentives could be provided for reducing lawn space, planting keystone species, or installing pollinator gardens. Municipalities could follow the lead of the San Antonio Water Utility, which is giving people 100-dollar coupons for adding natives to their yard. They could provide per-square-foot rebates for replacing lawn, the way California has done. These types of incentives provide immediate, short-term benefits, even if the benefits are small. But equally important, they change the psychological calculus by motivating homeowners to do something that is now "socially acceptable."

Are any native plants toxic to dogs?

The list of plants that could be toxic to dogs if they ate them is so long it's not worth printing. These plants include favorite non-natives as well, such as tulips, hyacinths, alliums, azaleas, and lilies. But they all taste bad, and it is unlikely your dog will eat them or eat enough of them to cause a problem. Dogs sometimes chew on plants when they are bored, or they may purposefully eat plants when they are nauseated to induce vomiting. But actual dog poisonings from eating plants are rare.

You often talk about bush honeysuckle when referencing the bad actors in the world of invasives. Are you referring to *Diervilla lonicera*? It is native to our Ontario landscape and we do sell it at our nursery.

Sorry for the confusion. I am referring to the Asiatic species *Lonicera maackii*, *L. marrowii*, and *L. tatarica*, all three of which are highly invasive, not to one of the *Diervilla* species that are native in eastern North America. That's one of the problems with using common names!

I notice in your books that jewelweed is not mentioned. I believe it is native and seems to be attractive to bees, insects, and even hummingbirds. I am wondering why this plant is not mentioned in your books?

You're right. Jewelweed is a great native plant, particularly for bumble bees and migrating hummingbirds. And it is a host plant for several caterpillar species. There is no particular reason why I haven't mentioned it. Nearly 3000 plant genera are native to the United States, including many that have excellent wildlife value. So, by the numbers, I don't mention most of them in my books—there just isn't enough space.

When buying native plants, how do I know if they have been treated with pesticides? What is wrong with purchasing native plants that have been treated with pesticides?

There is only one way to know for sure whether a plant has been treated with an insecticide, and that is to ask the nursery. If they are ethical, you will get an honest answer. Also ask which insecticide was used. A neonicotinoid insecticide is systemic, meaning it will be incorporated throughout the plant. Those insecticides can remain active for months in herbaceous plants and years in woody plants. Obviously, it is best to avoid plants treated with systemic neonicotinoids.

How do I find native planting resources in my area?

One of the reasons Michelle Alfandari and I started Homegrown National Park was to provide the resources you seek. Check HNP's website for a comprehensive resource of organizations and businesses supporting native plantings. These include national, regional, and state resources including nurseries, landscape designers, organizations, and more to help you get started with and maintain productive native plantings. Also, every state has a native plant society, and many of them can send you to local nurseries that sell natives.

Does poison ivy contribute ecologically? Everybody says to get rid of it.

Everybody except me! In terms of ecological function, poison ivy is a good plant. It makes berries in fall that are high in fat, which is just what migrating birds need. (Note: though good for birds, poison ivy berries are toxic to mammals—not that any wild mammal would ever eat them.) Its leaves support 19 species of specialist moths, it has great fall color, and it can climb trees without girdling them or pulling them down. Not only all that: bees love poison ivy pollen! In a study in Burlington, North Carolina, pollen brought back to honey bee hives was analyzed by its DNA to determine which plants the bees were foraging on the most. Pollen from 18 plant species was identified but poison ivy led the way, with a greater percentage of all the pollen collected coming from its flowers than from any of the other plants, which included white clover, honey locust, and dandelions, all supposed favorites of honey bees.

Yes, I get it that poison ivy oils usually cause blisters that really itch. I have gotten poison ivy every year of my life since I was six years old. The key is risk management. We all love beautiful roses, but if we were to grab a rose stem with bare hands, the sharp thorns would wreak painful havoc. But we know roses have thorns, so we carefully avoid them. Risk management—we do it every day without thinking about it. We just need to apply a little risk management to poison ivy. The vast majority of people who get poison ivy get it when they are trying to kill it! Leave it alone and you will be fine.

Where do you rank maples with regard to usefulness to wildlife?

In the mid-Atlantic States, maples are ranked seventh out of 830 plant genera in terms of supporting caterpillars. They also make copious amounts of seeds that are consumed by birds and mammals, and even though they are wind pollinated, many early season bees visit maple flowers for pollen and nectar. I'd say that's a pretty good ecological showing.

Here in southern Illinois, we enjoy the glossy, green leaves and lovely, white fragrant blossoms of our southern magnolia that was on our property when we moved in. It also provides a bit of privacy while we sit on our front porch. But how is that tree supporting our ecosystem? I am tempted to cut it down to allow one or more of the oak seedlings that volunteered beneath it to take over—should I?

Illinois is well out of the natural range of your magnolia, and even if it were within range, *Magnolia grandiflora* supports very little wildlife. That doesn't make it a bad plant. It's just not great at supporting wildlife. I suggest a compromise: enjoy your magnolia and the privacy it provides while you encourage your oaks elsewhere.

You frequently discuss the benefits of wild cherry trees. Do plums (*Prunus* spp.) differ dramatically as a host plant, or do they support a similar number of Lepidoptera species?

No one has carefully compared caterpillar use among all plum species, so in the absence of data, I will make an educated guess. Because plums and cherries are in the same genus, I'll bet the phytochemicals they employ to defend their leaves are very similar. And if that's the case, caterpillars adapted to those chemicals in other species of *Prunus* should be able to use plums quite well.

Oaks are not common in the foothills above Boulder, Colorado. We do have chokecherry, and I wonder if these are the most essential native trees in our ecosystems, especially since wildfires, dwarf mistletoe, and beetles are reducing the number of iconic ponderosa pines in some of our foothill areas?

The most productive plants in many areas of Colorado are riparian species that grow along water courses. Willows lead the way, supporting 328 caterpillar species. Cottonwoods are next with 269 species, and then your chokecherry with 259 species. Oaks, birches, and alder follow, and then pines. All are great plants, but to answer your question, you live in one of the few areas of the country where oaks are not the number-one producers of caterpillars, though they are still keystones.

I would like to have more shade and yet attract songbirds and eastern tiger swallowtails. I have considered planting a tulip tree, but I've learned they can grow to 200 feet and may be a hazard near (30 feet from) our house, and they break easily in high winds. What tree would you recommend?

Mature tulip trees (*Liriodendron* spp.) near the house can indeed hold your attention during high-wind events. They are known as the tallest, straightest trees in eastern forests. When they're surrounded by other trees, they are quite stable, but when we cut down surrounding trees and leave one or two isolated tulip trees, they have lost the buffering effect that neighboring trees supplied while they were attaining their great size. Almost any other tree species will be stable in a yard, especially if it grows in plenty of sunlight. Such trees grow out as much as they grow up and thus are harder for storm winds to blow over. A great way to stabilize the trees in your yard is to plant them small and in groves of three or four, close enough to one another that their roots can intertwine to form a very stable matrix. Trees anchored by a group root matrix are far more stable in high winds than isolated specimen trees.

I would like to change the landscape practices of the nursery I work for so that they use more natives in their designs. I haven't introduced to them the concept, but I appreciate your experience in approaching for-profit organizations with the idea of using native plants.

I appreciate your efforts to move the "green" industry more toward native inventory and truly "green" practices. I would not try to convince your boss using moral arguments or even ecological arguments. If your boss were sensitive to those approaches, key natives would already be used in the nursery's designs. But I *would* use economic arguments. Nursery owners are business people who make their living selling plants. I don't believe they care which plants they are selling, as long as sales are good. So I would point out that the market for native plants is finally in place. The demand exceeds the supply, which is exactly what anyone who has the ability to provide a product wants.

The native plant movement is not a fad that will soon disappear; it is the basis of a sustainable future, and an unsustainable future is not an option. Your boss can lead the cultural change toward ecologically productive landscaping (which is well underway) or let others lead and hope to catch up later. You might also point out that the old standby invasives—plants like Callery pear, burning bush, and barberry—are now banned in Delaware and other states, so it's a great time to build new inventory.

I read that red maple is the most common tree in the United States. Does it have the strongest chemical defense? How many insects feed on it?

You're right. Google says that red maple is now the most common tree in the country (and we all know Google is never wrong). But red maple has not always been the most common tree. Only after the loss of billions of ashes, hemlocks, American chestnuts, and American elms from invasive diseases and insects, and the reduction of oak populations in our forests by half from fire suppression and logging, does red maple increase in relative abundance. I can assure you that red maple is the most common tree not because it defends itself against caterpillar infestations better than other trees. In the United States, red maples support 363 species of caterpillars, making it one of our top 10 producers. Tulip trees support just a few dozen caterpillars, which means these trees are better defended from infestations, yet tulip trees are not our most common trees. And American yellowwood doesn't support any caterpillars, and it is even less common. Red maple is common today because it grows well in a number of soil, moisture, and light conditions and its major competitors have been reduced in number.

Cultivars of Native Plants

What is the difference between a variety and a cultivar?

A variety arises in nature and does not require human intervention to grow and reproduce, whereas a cultivar arises through human cultivation activities. Differences in terminology break down quickly when we find an attractive variety in nature and then clone it to sell as a named cultivar. The terms are used interchangeably by laymen and horticulturists, so, in practice, there really isn't much difference. Functionally, both cultivars and varieties are particular genotypes that express a desirable trait.

Blackberries are listed as native plants that support a high number of species. Does that mean that you could plant any type of blackberry? Specifically, I was wondering about some thornless varieties that have been developed.

The genus *Rubus* is a large one, and all of the native species support wildlife well. I don't think anyone has compared thornless cultivars with straight species, but my guess is they will support caterpillars just as well. Keep in mind that several non-native species of *Rubus* are now common throughout the United States as invasive species. They do not support caterpillars very well at all, so make sure you use a native species.

Are you opposed to cultivars?

Sometimes! It's not a black-and-white issue. The real question is, do cultivars function ecologically as well as straight species? The answer depends on which genetic trait was modified to create the cultivar. Some cultivars' traits (such as breeding to change green leaves into red or purple leaves, for instance) reduce insect use and therefore I oppose using them. Other traits, like making a tall plant short or introducing disease resistance, seem to have no impact on the insects that need those plants. In fact, such cultivars increase the likelihood that we will use them in our landscapes. One consistent downside to cultivars, though, is that they are clonally propagated—in other words, they have no genetic variation. And we know the loss of genetic variation in a population is not a good thing, particularly in the age of climate change.

Is there a concern that using cultivars may ruin the gene pool of the natives if there is successful cross-pollination?

I was part of a team of plant geneticists led by Andrea Kramer who tried to answer that question based on actual data. There are many warnings about what you describe in the literature, but there are no cases where it has actually been demonstrated. Typically, when a cultivar crosses with a straight species, the straight species wins and quickly overwhelms the cultivar's genome. Gene pool swamping appears to be one thing we needn't worry about.

Do you have a list of which native cultivars are OK? I am having trouble finding that information, and I don't want to place a cultivar in my garden if it doesn't offer optimum wildlife value.

No such list exists. Thousands of cultivars are available, and no one has researched all of them. Two generalities might help, though: Avoid double flowers because they are sterile and offer no pollen or nectar to pollinators, and avoid red-leaved cultivars because they are distasteful to insects. Mt. Cuba Center in Hockessin, Delaware, has been doing some excellent trials with native cultivars, so check the website for the latest results.

I have a cultivar of switchgrass called 'Shenandoah', which turns red in autumn and then tan in winter. It is not red in spring or summer when most insects are looking for a host. Is this still a host plant? I want to encourage egg laying by insects that use grasses.

No one has compared host use of switchgrass cultivars with that of the straight species. But my guess is your cultivar will perform as well as the straight species during the growing season when caterpillars (many skippers and others) are developing, because the red color (anthocyanins) hasn't yet built up in the leaves. Anthocyanins discourage caterpillar and grasshopper use.

Keystone Plants

What is a keystone plant and how do you determine that?

The term *keystone* refers to the stone in the center of a Roman arch; the arch depends on that stone to stand, so if you remove it, the arch will collapse. The concept of a keystone species was first published by ecologist Robert Paine in 1969, when he was studying the role of starfish in tidal pools along the Pacific Coast. He found that if a starfish was removed from a tidal pool, the species inhabiting the pool after the starfish was removed changed—that is, the ecosystem in that tidal pool ceased to exist as it was after the main predator, the starfish, was removed. I extended the use of the term to plants in a paper with Desirée Narango and a research assistant in 2020. A keystone plant performs an ecological function much better than most other plants. My students and I, as well as Jarrod Fowler of Pollinator Partnership, have focused on two categories to designate a keystone species: how well plants support caterpillars and how well they support pollinators (such as bees). We are interested in plants that support pollinators for obvious reasons: Pollinators enable 80 percent of all plants and 90 percent of all flowering plants to reproduce. Plants that provide the pollen necessary for most pollinators to reproduce are therefore essential to the continuation of terrestrial ecosystems.

The reason we *should* be interested in plants that support the most caterpillars is not so obvious. It turns out that caterpillars serve as a vital link between plants and the food webs for which they supply energy. Plants capture energy from the sun and, through photosynthesis, turn it into simple sugars and carbohydrates, the food that supports just about all of the animal life on Earth. Way back in 1988, Dan Janzen estimated that plant-eating caterpillars transfer more energy from plants to other animals than any other plant eater. This makes plants that host many species of caterpillars the keystones in terrestrial ecosystems. If you remove those plants, food webs collapse, just the way a Roman arch collapses if you remove the keystone at the top of the arch. We learned that just 14 percent of North American native plant species are supporting 90 percent of the caterpillars that feed other animals. Those are the species I call keystone plants. We determine which plants are keystone species through host records published in the literature over the last century.

In a given landscape, is there an ideal percentage of biomass that should be made up of keystone species?

That's like asking what the ideal size of a bank account is! The more the better! Every landscape needs keystone plants, needs to sequester carbon, and needs to do the best it can to support pollinators and manage the watershed. The balance will depend on many factors, especially space and sun availability. You don't want a diversity of native species that are less productive. Diversity is good as long as a landscape is well endowed with keystone species. Remember that plants that support lots of specialist pollinators are often not the same plants that support lots of caterpillars (although some genera, such as *Solidago*, *Symphoritricum*, and *Helianthus* do both!), so if the landscape meets the needs of caterpillars and pollinators, it will have all the diversity it needs.

What should we do with non-native species within keystone genera— flowering cherry and other non-native *Prunus* species? Weeping willow (*Salix babylonica*)? Non-native *Pinus* species? Are these not considered keystone species in the United States? I imagine the answer varies between plant species and the ecoregions they are grown in.

I asked that question with my student Karin Burghardt in a big, five-year common garden experiment (all treatments grown in the same garden), which compared insect use of native and non-native congeners (plants within the same genus). On average, insect use, particularly caterpillar use, of non-native congeners was reduced 68 percent compared to use of native members of the genus. This did vary depending on the genus in question, but the safest bet is to use the native species in a keystone genus when there is a good alternative to a popular non-native ornamental in that genus. Otherwise, the keystone value of the genus is largely lost.

Are there keystone conifers?

Yes, indeed, but these differ as you move around the country. Native pines are keystone species wherever they occur. In the northern parts of the United States, spruce, larch, and firs are keystones species, although they do not function as keystones when planted at latitudes south of their natural ranges.

Chestnut doesn't rank highly on the Native Plant Finder in terms of Lepidoptera usage, but I'm sure you're aware that it *is* strong in terms of other functions of keystone species, such as providing wildlife food. Its consistency in year-to-year nut production is much better than that of the other seed producers in the forest, and the mast crop, of course, is massive. So it's not great for Lepidoptera, but it is great for many other types of wildlife that eat nuts. Would you not consider the American chestnut a keystone species?

We can define keystone species in terms of ecological roles. What is listed on NWF's Native Plant Finder are the keystone species defined by their production of the caterpillars that are so important to most terrestrial food webs. The plant finder's database was created from 4000 literature citations made over the last century of Lepidoptera host plant records. Chestnut isn't high on the list because they are essentially gone. They are in the family Fagaceae, and, as close relatives of oaks, there is no doubt in my mind that they would support many caterpillar species if they were still common trees. Their absence from our lists is surely an artifact of the chestnut blight.

In your book *Nature's Best Hope*, *Solidago* is considered a keystone genus. In your book *The Living Landscape*, you say that it doesn't seem to be as robust as many trees. Does the keystone plant concept differ between woody and herbaceous plants?

If we compare all woody plants to all herbaceous plants in terms of their ability to support caterpillars, the woodies, on average, support more caterpillar species. But if we look *within* woodies or herbaceous plants, the same pattern exists: just a few of the plant genera are supporting most of the caterpillars. Goldenrod (*Solidago* spp.) is ranked near the top of the list of herbaceous plants across the country with regard to its ability to support caterpillars. So we can call them "keystone herbaceous plants." In Chester County, Pennsylvania, for example, goldenrod supports 110 species of caterpillars. That's not as good as oaks, which support 511 species, but it is still better than any other herbaceous plant genus. If we measure plants in terms of the number of specialist pollinator species they support, *Solidago* species are again excellent keystone plants, surpassing all woody plants. So when talking about keystone plants, we need to specify whether we are talking about woody plants or herbaceous species, and we need to indicate the function we are referring to: making caterpillars, supporting pollinators, or some other ecological function.

What plants may be causing harm to the (our) environment or just taking up space? My acre yard has mostly non-native trees and shrubs. I've taken out the most invasive plants, but is there harm in keeping those that are non-native but not invasive?

The "harm" you refer to is a matter of scale. A non-native plant that serves primarily as a decoration is like a statue: it takes up space that could be used by a productive plant, but a few statues here and there are easy to accommodate in a well-designed landscape. If most of your landscape comprises decorative plants with little ecological value, however, the amount of "harm" to your local ecosystem increases.

I have around 5000 square feet of garden space. I have been replacing non-natives with natives for three years. Do any non-natives provide some value?

In terms of wildlife value, let's group plants into two categories: those whose primary value is for flower visitors, including pollinators; and those that convert energy into insect food, especially caterpillars and orthopterans such as grasshoppers, katydids, and crickets, which feed the birds, foxes, shrews, bears, and other animals. Most flower visitors are generalists and are seeking nectar and, to a lesser extent, pollen. Many non-natives that are not invasive are rich in nectar (zinnias, marigolds, white clover, English lavender, and Russian sage come to mind) and meet the needs of generalists while they are in bloom. Although many invasive plants are visited by generalist insects during the brief period they are in bloom (such as privet, Japanese knotweed, autumn and Russian olive, buckthorn, and Callery pear), under no circumstances should you consider any of these plants good for pollinators. Because they eliminate a diverse community of native flowering plants when they invade a habitat, invasive plants hurt flower visitors far more than they help them. Also keep in mind that non-native plants are not meeting the needs of specialist pollinators that can rear young only on the pollen of plants they have coevolved with over the eons. But building a garden with both essential natives and some nectar-rich non-natives that are not invasive is fine.

A few non-natives also support leaf-eating insects, though not very many. Leaf eaters are among our most specialized insects and typically cannot use plants they did not coevolve with. Exceptions include non-native willows, apples, and members of the carrot family such as parsley and dill. When a non-native is closely related to a native plant and shares the same leaf chemistry, some of our native insects can develop successfully on their leaves.

CHAPTER FOUR

Oaks

Where do oaks occur in the United States?

Oaks are one of the most ubiquitous tree groups in the country. We have ninety-one species of oaks in the United States, and at least one species occurs naturally in all states except Alaska, Hawaii, Idaho, and Montana. They also occur only in the south-central and northeast parts of Wyoming and the mountains of southern Nevada. The Southeast supports the most oak species.

Under optimal conditions, how long can an oak live?

The average life span of an oak is around 900 years, much longer than you might think. Some individuals live much longer. Many oaks throughout the country have exceeded 500 years in age (such as the Angel Oak in South Carolina and the Seven Sisters Oak in Mandeville, Louisiana). The Pechanga Great Oak, a coastal live oak in Temecula, California, may be 2500 years old, but, surprisingly, small, ground-hugging species such as the Palmer oak (*Quercus palmeri*) can be some of the oldest living things on the planet—up to 13,000 years old. Most oaks in our neighborhoods don't reach those ripe old ages because our land management practices do more to kill the oaks than protect them. We habitually rake away all of their fallen leaves, preventing the return of the nutrients in those leaves to the soil so our oaks can reuse them in the future. We are pruning fanatics who remove branches that do not need to be removed, and we often prune during the warm seasons, giving pathogens access to oaks' vascular systems. Moreover, our roads and pipelines and sewer systems prevent oak roots from growing to their optimal length.

Slow-growing oaks like the Palmer oak can live for many thousands of years.

Is it true that most oaks are planted by squirrels and Blue Jays?

I would say most oaks are planted by jays (of all species), with a few being planted by squirrels. A jay can bury up to 4500 acorns each fall, but it remembers the locations of only 1 out of every 4 acorns it buries. This means that in a good year, a single jay can plant more than 3300 oak trees! Jays can carry acorns up to a mile (some say a mile and a half) before burying them, so each acorn is well removed from competition with its parent tree. Squirrels, in contrast, eat far more acorns than they bury, they tend to cache the ones they do bury instead of burying each one individually, and the distance they move an acorn is measured in yards, not miles. Some gray squirrels learn to snip off the germinating end of an acorn from the white oak group so that it won't continue to shunt energy into an elongated root. That interesting behavior preserves more nutrition in the acorn itself—good for the squirrel—but it ends the potential for the acorn to grow into a tree.

Jays of all species, like this Western Scrub-Jay, bury acorns in fall to store for winter food.

How do jays extract acorns from frozen ground?

Jays have powerful, all-purpose, sharply pointed beaks. I have never witnessed it, but I can easily imagine them using that beak as an ice pick, chipping away at the substrate until the acorn is free.

How important are the caterpillars on oak trees?

They are enormously important, at least from an ecological perspective. Caterpillars are the bread and butter of temperate zone terrestrial food webs, and they are very important in the tropics as well. They transfer more energy from plants to other animals than any other plant eaters. Remember that plants capture energy from the sun and, through photosynthesis, convert it to simple sugars and carbohydrates—the food that supports just about all animals on the planet. But that food supports animals only if it reaches the animals. Most animals do not eat plants directly; they eat insects that eat plants, and

Caterpillars are the bread and butter of terrestrial food webs, feeding birds, mammals, and reptiles like this ornate box turtle.

most of those insects are caterpillars. Without caterpillars, the vast majority of our birds would not be able to reproduce, and countless lesser-known animals would disappear as well. This is why oaks are so important compared to other plants; they support the growth and reproduction of the most caterpillars!

I live outside of Seattle and I am trying to encourage Garry oaks on my property. Douglas firs are seeding in and shading out my young oaks. Should I cut the firs down?

Yes, you should. The oak savanna ecosystem of the Pacific Northwest is one of the most imperiled ecosystems in North America. Nearly 99 percent of these ecosystems have been destroyed by agriculture and development, endangering many animals and plants that depend on this fragile ecological association. With so few remaining oaks to save, efforts are now focusing on restoration, planting young oaks just as you are doing and managing the grasslands among

Even though oaks are wind pollinated, many native bees use oak pollen to rear their broods in spring.

the oaks with fire. The biggest threat to these restoration projects is the incursion of Douglas fir. So even though Douglas fir is an important native tree, we need to manage it or we will lose forever the oak savannah ecosystem and the diversity associated with it. From the perspective of caterpillars (bird food) alone, Garry oaks are superior to Douglas firs and support 185 species of caterpillars, while Douglas firs support 130 species.

I know that oaks support more species of caterpillars than most/all other species of trees. However, I do not know anything about the abundance of caterpillars on other species of trees. Is there evidence that oaks produce more caterpillars per tree or per unit of biomass than other tree species?

The evidence that oaks support more caterpillar species than other trees comes from host records over the last century. When a scientist finds a caterpillar eating a particular tree and then goes to the trouble of publishing that observation, it becomes a permanent host record. It becomes a verified host record when several other people publish a similar observation, suggesting that the first observation wasn't a case of mistaken identity. You can find out how many caterpillar species are supported by various tree species in your county by visiting the Native Plant Finder online. Oaks support more species than other trees in 84 percent of the counties in which they occur naturally. But native cherries and plums, native willows, native poplars, native birches, and native maples all support hundreds of species as well. Unfortunately, no one has tested whether caterpillar abundance is correlated with species richness. That's on my to-do list.

I planted hundreds of germinated acorns on my land in Wisconsin. I'm wondering if squirrels like germinated acorns as much as ungerminated ones, and how many they will dig up?

Squirrels prefer ungerminated white oak acorns. A squirrel can even learn to snip off the end of an acorn so that it will never germinate, which enables it to store the acorn for months. But they will also eat germinated acorns, especially if other foods are in short supply, and they will even dig up seedlings the following spring to eat what's left of the acorns that are still attached. One sure way to get a tree out of an acorn (and avoid it becoming a meal) is to cage it—build a small wire cage and place it around the planting spot to keep the animals out; this will protect it from not only squirrels, but from mice, voles, and deer as well. It's worth the trouble!

You say that oaks keep carbon out of circulation longer than shorter lived pines and poplars. When a pine is harvested for its wood and that

wood is used for home or furniture construction, does the carbon it has sequestered remain in the wood as long as the home or furniture exists?

Yes, it does, but that goes for oak furniture as well! The wood used to build George Washington's Mount Vernon 250 years ago is still holding all of the carbon it was made from. And if much of the wood came from mature trees, carbon may have been kept out of the atmosphere for 500 or more years.

I planted a white oak tree years ago. I see green acorns growing, but we have never seen one on the ground, ever. The chipmunks and squirrels and other wildlife must get them all. When are the acorns mature enough to pick and plant?

Acorns are not like peaches. You don't pick them; you wait until they drop of their own accord. And most of the animals that eat acorns do not pick green ones off the tree. The fact that you never see mature acorns on the ground makes me think those acorns were not fertilized. Unfertilized acorns are aborted by the tree way before they mature. If your oak is the only oak of that species in the area and is outside of the pollen dispersal range from other oaks of the same species, the acorns will not be fertilized and will be aborted.

Why do fallen oak leaves take so long to break down?

Oak leaves are chock-full of lignans and tannins, which make them tough and rot resistant. Only after most of the tannins have leached out of the leaves can bacteria, fungi, and insect detritivores attack them in earnest.

What are the water requirements for oaks?

Water requirements depend on the oak species. Pin oaks are also known as swamp oaks, and along with swamp white oaks, they do well in moist bottomland. Chestnut oaks and Georgia oaks, in contrast, prefer dry, rocky outcrops and will object to too much water. Many oaks in the dryer areas of California need water during winter months but can be killed by supplemental watering in the summer months. Learn what species of oaks you are managing and ask the Google for watering requirements specific to that species. My rule of thumb is this: once your oak is established, it should not need any supplemental watering if you have chosen a species appropriate for your ecoregion and soil type.

Which oak grows the fastest?

Under the right conditions, most oaks grow considerably faster than most people think, but if we were to have a race between members of the red oak group and white oak group, the red oak group would win. Northern red oak, in particular, grows fast in rich, deep soil and normal rainfall. Twenty-one years ago, I planted a bunch of acorns to create an allée of red oaks along my driveway. The soil is not very rich and was heavily compacted by the company that installed my driveway. Nevertheless, those trees are now 60-plus feet tall!

Do all oaks produce acorns, or are the trees only male or only female?

All oaks are monoecious, meaning each individual tree has both male and female flowers. Female flowers are tiny green structures that we have to look hard to find; if pollinated, these are the structures from which acorns grow. But all oak individuals also produce catkins, male flowers that release huge loads of pollen on the wind. Many monoecious plants can self-fertilize, but not oaks. In oaks, the female flowers mature on an individual just after the male flowers release their pollen. Thus, female flowers on an oak must be fertilized by the pollen of another oak tree nearby whose pollen was released at the same time the female flowers on the first oak matured.

This slight asynchrony among individual oaks has important ramifications for how we plant them if we want our oaks to produce acorns. If we plant only one individual of an oak species and no other members of that species are within pollen dispersal range (maybe 100 yards on a good day), our tree will never make acorns. I watch this reproductive failure unfold in my yard every year. Shortly after we moved in, I planted a (now very large) shingle oak, but the next nearest shingle oak is many miles away. Every year, my shingle oak sets thousands of acorn embryos, and every year it aborts every one of them because they were never fertilized.

Does anything eat red oak scale?

You are probably referring to the kermes scale, an insect that can build up to great numbers on red oaks under the right conditions. Yes, there are predators

An oak produces both male catkins (the long, dangling green objects) and female flowers (shown here as tiny pink flowers) on the same individual.

The caterpillar of the beautiful kermes scale moth (shown here) is unusual because it is a predator on kermes scale.

of scales, including most species of coccinellid ladybird beetles and the kermes scale moth, which specializes on that particular creature. The caterpillar of this beautiful moth devours kermes scales from May through September.

In general, how does insect activity on oaks vary based on the age of the tree?

Though this has never been formally studied, I can tell you that insects, particularly caterpillars, use oaks from the time the trees germinate until well after they die. It is easy to imagine that the larger the tree, the more food it provides for insects, so insect abundance on oaks almost certainly increases with oak age.

What mushrooms should I plant in my New Jersey native garden to enhance oak trees' nurturing capabilities via mycorrhizae?

You cannot plant mycorrhizal fungi, but you can encourage the soil conditions necessary to sustain them. First and foremost, do not apply any pesticides. Remember that many lawn fertilizers contain broadleaf herbicides designed to kill everything except grass. If you have grass under your oak and put down fertilizer, you will harm the mycorrhizal community in the soil. Next, you want to build up the organic matter in the soil. You should do this naturally by allowing the leaves that fall from your tree to remain under your tree. As the leaves break down, they add to the soil the organic matter needed by mycorrhizae and your tree to thrive. You do not have to add mycorrhizae to your soil. Their spores are floating on the wind all the time and will colonize the soil under the tree if it is rich in organic matter.

What birds rely on oaks for food and shelter?

Just about any bird that relies on caterpillars to feed its young relies on oaks. And that's a lot of birds. Most North American terrestrial bird species (96 percent of them) rear their young on insects, and most of those insects are caterpillars. So it stands to reason that the tree species that produce the most caterpillars, and oaks lead in that category, would be the most valuable to those birds. Many birds also rely on the acorns that oaks produce to get them through winter, including all jay species, Acorn Woodpeckers,

Red-bellied Woodpeckers, flickers, towhees, titmice, Wood Ducks, turkeys, and even Sandhill Cranes. Oak snags also help provide homes for the 85 species of North American birds that nest in tree holes. No other tree genus does more for our birds throughout the year.

Why are oak acorns not able to be "seed saved"?

For one simple reason: most oaks germinate the same year they mature as a seed. Most members of the white oak group germinate within days of falling to the ground. Red oak group acorns require a dormancy period that includes exposure to winter cold temperatures; then they germinate in spring. So even though long-term storage is not an option, red oak group acorns can be stored in a refrigerator during winter and then planted in spring.

Do oaks have medicinal value?

The ancients would have said that oaks definitely have medicinal value! Oak parts, particularly bark, were used extensively by Indigenous groups throughout oak ranges. Though I cannot verify the effectiveness of any of the claims in the literature, the high tannin content of oak bark made it valuable to Native peoples for its astringent qualities. Concoctions made from oak bark were used to treat diarrhea, colds, fever, coughs, bronchitis, burns, gum disease, and as a general antiseptic.

Can you speak to the importance of oak savannas? We have a local treasure in a remnant tallgrass prairie and black soil–bur oak savanna and would love to hear your thoughts on the importance of conserving habitats like this. Some of the oaks are 200 years old.

Oak savannas were good places for humans to settle because they were so easy to destroy, which is why so few of these unique ecosystems remain. Only 0.5 percent of the oak savannas that once made up a transition ecosystem from eastern deciduous forest to treeless prairie, ranging from Canada to Texas, remain. Oak savannas are one of the most endangered ecosystems on the planet. These communities are maintained historically by several types of disturbances, including fires set naturally by lightning and later by

Indigenous humans, grazing by millions of bison and elk, low precipitation, and poor soil. Only trees with a high tolerance for fire, primarily bur oak (rich soils) or black oak (sandy soils), were able to survive this fire regime. Oak savannas are one of our most diverse ecosystems because they combine the productivity of oaks with the rich productivity of prairie plants in ways that no other ecosystems can.

Are any insects detrimental to oaks?

Yes, and there are many. In fact, thousands of insect species depend on oaks for nutrition, but very few weaken oaks enough to shorten the trees' life spans. Oak twig borers, weevils, bess beetles, metallic wood-boring beetles, lace bugs, several species of scales, more than 1100 species of caterpillars, several species of bark beetles, more than 300 species of gall wasps, sawflies, assorted aphids, leafhoppers, planthoppers, treehoppers, katydids, walking-sticks, tree crickets, and other insects regularly live on oaks in North America, but their impact is benign, and only rarely are they even noticed. The insects most likely to hurt oaks are invasive species from other continents, and at the top of this list is the spongy moth. Brought to this country from Europe in the late 1800s in a misguided attempt to breed a super-duper silk moth, the spongy moth has become invasive and has degraded oak forests of the Northeast. And this insect is a constant threat to the magnificent forests of the Ozarks. In addition, the goldspotted oak borer, a wood-boring beetle native to Arizona, has moved (been moved?) into California, where it is not controlled by natural enemies and is threatening many of the oak species in that state. Several species of bark beetles are often reported as pests of oaks, but they attack only oaks that have been weakened by disease or some other environmental trauma and are not a threat to healthy oaks.

Do oak forests depend on fire?

Yes they do, at least to some degree. Young oaks growing in the shady conditions of a forest grow quite slowly and are readily shaded out by fast-growing cherries, birches, and maples. Regular ground fires, however, keep those competitors for light at bay. Oaks have comparatively thick bark that easily protects their cambium from the deadly heat of ground fires. Not so cherries, maples, birches, tulip poplars, and other more vulnerable tree species. Native peoples knew this and regularly set ground fires to manage forests for oak dominance. Not only did these savanna-like open forests favor oaks over other species, they provided Indigenous people with acorns, an important source of nutrition that also nourished deer, turkeys, squirrels, and black bears—all of which were game species for Native Americans. When Europeans displaced Native Americans and suppressed fire, oaks started a slow decline in dominance in our forests. Today in North America, there are 50 percent fewer oaks than there were 200 years ago.

What role does fire have in acorn germination?

Unlike many pines, oaks do not need fire to trigger germination in their seeds (acorns). In fact, fire kills acorns, so prescribed burns should be conducted before acorns fall or in years when there are very few acorns. You wouldn't want to do a prescribed burn during a mast year. That would retard oak reproduction for years.

Which oak has the smallest acorns?

The smallest acorns are produced by the laurel oak, though willow oak and water oak acorns are also quite small. You didn't ask, but the largest acorns are made by bur oaks.

How much space do you need to grow oak trees?

The answer depends on the species you choose and the amount of sun it is exposed to. Most oak species grow into a wide-spreading habit in full sun; in fact, mature trees can have a canopy spread wider than the height of the tree. I have a shingle oak in my yard that has grown rapidly in the last 20 years and now occupies a space about 60 feet wide. White oaks that I planted as acorns 21 years ago have now spread to 45 feet wide. Of course, the small oak species in my yard, such as dwarf chestnut oak, have reached less than half that size in the same time period.

I have four oak trees around my home. Three of them have trunks that look as if they were two that became one. I have noticed other oaks that look this way. Could you talk about branching in oak trees? Is this decurrent branching? Or have two trees merged?

My guess is that, rather than decurrent branching (weak apical dominance), this is the result of the leader being eaten or cut off when the tree was small. Deer browse often results in a similar branching pattern, but so does coppicing.

How important are oaks to butterfly life cycles?

Oaks are wonderful host plants—in fact, oaks are the best host plants for moths in North America because they support more than 1100 species. For some reason, far fewer butterfly species use oaks to nourish their caterpillars, but several common butterflies do specialize on oaks. In the West, the beautiful California sister develops on oaks as do many duskywing species. Several species of hairstreaks also specialize on oaks in the West, including the golden hairstreak, gold-hunter's hairstreak, and California hairstreak. In the East, other hairstreaks and duskywings depend on oaks, including the striped hairstreak, banded hairstreak, Edwards' hairstreak, red-banded hairstreak, white M hairstreak, Horace's duskywing, and Juvenal's duskywing.

I have heard you compare the number of caterpillar species found on oaks to the number of species found on ginkgo trees, but that seems inappropriate. It is comparing a genus to a single species. Please provide information on specific oak species relative to other species.

In other words, if I searched for caterpillars on ginkgoes and just one species of oak, would I find the same number of caterpillars on both trees? No, I wouldn't. First, let me explain why we make comparisons at the genus level rather than the species level. Because most host records in the literature are at the genus level (eats oaks) rather than the species level (eats white oak), species records are woefully incomplete. But even given this inaccuracy, literature host records and any field experience will quickly convince you that oaks support more species than ginkgoes or any other non-native tree.

Juvenal's duskywing caterpillars specialize on oaks over much of their range.

Even though host records are incomplete for oak species, records show that northern white oak supports 283 species of caterpillars, Oregon white oak supports 110 species, bur oak supports 102, pin oak supports 59, northern red oak supports 213, and so on. So at the genus or species level, oaks contribute orders of magnitude more energy to local food webs than do ginkgoes, which support no caterpillars.

Another way to look at it would be to compare native tree genera with only one species and to ginkgoes. The tulip tree, for example, is the only species in its genus (*Liriodendron*) in North America, yet it supports 26 species of caterpillars, while ginkgoes support none. Does a bird seeking caterpillars care if the comparison is appropriate or not? If you are a hungry bird looking for caterpillars, would you go to an oak or a ginkgo? If I offered you 1000 dollars for every caterpillar you could find in 30 minutes, would you spend that precious time searching ginkgoes or oaks? It may be an "inappropriate" comparison, but I know where you would search.

I think most of your research on how important oak trees are to the ecosystem has been done on the East Coast of the United States. I would like to know how applicable that research is to other parts of the country. I never see many caterpillars on our native Gambel oaks and wonder about the research.

We have caterpillar host records for oaks in every county of the country in which they occur nationwide. Gambel oaks support many caterpillar species—221 species to be exact. Try looking for caterpillars on your oaks at night. They are not hiding from the birds then and are easier to find.

When using your statistics about the keystone status of oaks and the diversity and sheer numbers of insects they support in my educational writing as a naturalist, I have gotten some pushback about whether your work on the East Coast is fully relevant for the Midwest (Minnesota). Any suggestions?

It's true that when I give talks, most of the photos I show to illustrate my various points are from the East. In fact, most are from my yard in Pennsylvania. But the research coming out of my lab at the University of Delaware is relevant nationwide, and the ecological principles revealed by this research are relevant globally. Tell people to check for themselves. Go to the online Native Plant Finder and compare the rankings of various plant genera in the Midwest with those rankings in the East. This is possible because we have ranked the ecological value of every plant genus in every county of the United States. Anyone who goes to the trouble of doing this will see that the principles are the same: everywhere you go, just 14 percent of our native plants support 90 percent of the caterpillars that drive local food webs. These are the keystone species I talk about so much, and they are surprisingly similar across the country.

I have black oaks at my house on a bluff on Lake Michigan. Pure sand for soil. I have only black oaks; the reds and whites have been used for lumber. Are black oaks of any value to wildlife? A forester told a group I was in that black oaks are useless.

I'm sure the forester was referring to the number of board feet of lumber typically produced by black oaks in sandy soil. Ecologically, black oaks are just as valuable as any other oak species.

What is the role of mycorrhizae in oak savannas?

Mycorrhizal fungi play an important role in moving nutrients through soil into plant roots, where they can be used for plant growth and reproduction. This is critically important in nutrient-poor soils. Oaks, like many other plants, have mycorrhizal associations that improve their growth and ability to defend themselves from insect herbivores. In a paper published in 2009, I. A. Dickie and other researchers compared the mycorrhizal community associated with oaks in oak savannas and oak forests. They found a similar diversity of fungal mycorrhizae in oak forests and savannas, but oak savannas housed unique mycorrhizal species. From a conservation perspective, this is worth noting, since oak savannas are one of our most endangered ecosystems, with only 0.02 percent of their original area remaining worldwide.

Is there a difference in nutritional value between acorns of different species?

As far as I know, the nutritional value of acorns from different oak species is about the same. What varies is the size of the acorn—the amount of food it offers—and the tannin content—how bitter it is. In general, members of the red oak group (red oak, black oak, pin oak, shingle oak, and so on) have higher tannin content than members of the white oak group (white oak, Garry oak, bur oak, post oak, swamp white oak, chestnut oak, and so on). If I had to choose one species that offered the best source of food for wildlife, I would choose the bur oak, because it produces the sweetest, largest acorns in the country.

I have several oak trees on my property. Is there an easy way to identify which is which? How many species are native here in Massachusetts?

Ten oak species are commonly found in Massachusetts: white oak, swamp white oak, scarlet oak, chestnut oak, chinkapin oak, pin oak, dwarf chestnut oak, red oak, post oak, and black oak. I suggest you consult the 2004 book *Native Trees for North American Landscapes*, coauthored by Guy Sternberg, to help identify which species you have. Or take a picture of one or more leaves of oaks that confuse you and post them on the iNaturalist app; within minutes, someone will identify the tree. A third option is to use the Seek app by iNaturalist: open the app, point the Seek Camera over a leaf, and you will get a reasonable guess about its species in seconds.

We have scrub oaks in Florida. Is the scrub oak the same species in other parts of the country and other eco zones?

Your question illustrates a general problem with using common names to identify plants: these names are often regionalized, and they can refer to different species in different regions. Pretty much any oak that is low in stature is called a scrub oak. California alone has five species of scrub oak: California scrub oak (*Quercus berberidifolia*), coastal scrub oak (*Q. dumosa*), leather oak (*Q. durata*), Tucker's oak (*Q. john-tuckeri*), and island scrub oak (*Q. pacifica*). In the Northeast, bear oak (*Q. ilicifolia*) and dwarf chestnut oak (*Q. prinoides*) are often called scrub oaks. In the Southeast, including Florida, no less than five oak species are called scrub oaks, including Chapman oak (*Q. chapmanii*), myrtle oak (*Q. myrtifolia*), sandhill oak (*Q. inopina*), sand live oak (*Q. geminata*), and turkey oak (*Q. cerris*). And in the Southwest, Coahuila scrub oak (*Q. intricata*), Gambel oak (*Q. gambelii*), gray oak (*Q. grisea*), Emory oak (*Q. emoryi*), pungent oak (*Q. pungens*), and Sonoran scrub oak (*Q. turbinella*) are all called scrub oaks by locals.

Why is oak used in making bourbon barrels? The bourbon industry in Kentucky is aggressively logging the white oaks.

It is not just bourbon and other whiskey producers who use oak barrels, but many wines are also aged in oak, particularly white oak. Why oak? Because oak wood influences the taste of the product within. But it is tricky business. With whiskeys like bourbon, the type of oak, the way in which the barrel is charred, and where the barrel is stored all influence the taste. For wine, slow aging in oak infuses the wine with oxygen, which softens its flavor and adds tannins that are said to add "structural qualities" (whatever they are) to the taste. This centuries-old dependency on oak, however, is concerning for the few white oak forests that remain in Kentucky, where 95 percent of the world's bourbon is made. Old-growth oak forests are being felled far faster than they are being replaced. With our ever-expanding demand for wine and whiskey, I cannot think of a way to make this industry sustainable, short of abandoning oak altogether and relying on chemical manipulations by talented food scientists to replicate the taste that years in an oak barrel produces. I'm sure this is possible, but many would consider it blasphemy.

How old do my oaks have to be before they make acorns?

Most species will not make acorns until they are at least 20 years old. I have found that smaller species, such as dwarf chestnut oak, make acorns much sooner—some after only eight, nine, or ten years. I have also been asked whether coppiced oaks will make acorns. The answer is no; the regrowth on a copse is all young wood incapable of producing flowers or catkins.

Is poison oak related to regular oaks?

No, the two are not even distant relatives. Oaks are trees in the genus *Quercus*. Poison oak is a vine in the genus *Toxicodendron*, and I have no idea why it is called poison oak. Maybe we should start a campaign to change the name?

Are willow oaks as useful as other oak trees?

Yes, if they are planted within their natural range, which is from south coastal New Jersey to northern Florida, and west to eastern Oklahoma. They are often planted outside of that range, but the creatures that use them usually don't occur in such places, and this reduces willow oaks' value to wildlife.

Why do bur oak acorns have such a thick coating?

The thick coating is correlated with the size of the acorn and where they grow. Bur oaks produce the largest acorns of any North American oak. They are also a fire climax species that evolved in oak savannas, where ground fires used to be common. The thick wall of the acorn and its very large and heavy cap may be adaptations to keep the seed safe from the heat of a wildfire. But why so big? One leading hypothesis is that the large size of bur oak acorns is an adaptation to the poor soils of areas that were glaciated. A big acorn can supply the seedling it produces with enough nutrients to grow a root system capable of extracting scarce nutrients from the soil. This hypothesis is supported by the fact that bur oak acorns are smaller in trees growing south of glaciated soils. Another hypothesis is that giant acorns are needed to produce seedlings large enough to compete with the dense prairie plant communities into which they fall. These two hypotheses are not mutually exclusive; they both probably have influenced the size of bur oak acorns over the course of their evolution.

Why do oaks support so many species?

This is a good question that no one has definitively answered, but at least six hypotheses have been proposed in research papers by leading scientists including Janzen (1968), Southwood (1983), Condon (2008), and Grandez-Rios (2015) that suggest possible explanations. When we say more caterpillars use oaks than other plants, we mean that more caterpillar species have adapted to the many phenolic compounds that characterize oak chemical defenses. So any feature of oaks that encourages such adaptations may help explain why so many species can now use oaks as host plants.

One such feature may be the size of the genus *Quercus*. There are about 435 oak species world-wide, with 91 occurring in North America, making *Quercus* the largest tree genus within the northern hemisphere. By comparison, there are 400 species of *Prunus* (cherries) and 350 species of *Salix* (willows) worldwide; these are large genera to be sure, but they still contain fewer species than *Quercus*. And tree genera such as maples (160 species), pines (111 species), birches (30–60 species), poplars (30 species), and tulip trees (only 2 species) are much more representative of average genus size among most trees.

Highly correlated with genus size is the geographic area covered by a genus. Tree genera that occur over large areas of the country—or, indeed, the world—overlap the ranges of many more caterpillar species than tree genera that are not widespread. This puts hundreds of caterpillar species in contact with widespread tree species for long periods of time, a prerequisite for the evolutionary interactions that lead to host use by caterpillars. Oaks occur across Asia and Europe, throughout the United States and Canada, and even throughout Central America and into the northern portions of South America, which constitutes the greatest geographic range of any tree genus.

Plant apparency (the likelihood of a plant being found by herbivores) has also been considered a factor in the evolution of host use. Plants that are large and/or present in the landscape for long periods of time are more easily encountered by moths and butterflies (caterpillar adults), and therefore they are more likely to develop close relationships with Lepidoptera species than tiny plants with very short life cycles. An oak, for example, that exists in exactly the same spot for hundreds of years and becomes an enormous individual over that time span is easy to find by a female moth loaded with eggs. It is far more likely that local Lepidoptera would adapt to oak tree defenses than to tiny spring ephemerals whose total biomass is less than that of a single oak leaf and whose entire life cycles are completed in a few weeks. Finally, the type of chemical defenses used by plant lineages has been implicated in influencing the rate at which insects adapt to plants.

Defensive chemicals have been classified into two groups: quantitative defenses and qualitative defenses. Quantitative defenses are chemicals that are not immediately toxic to consumers but become more effective through repeated exposure—that is, they work best as they accumulate in a caterpillar's body. The tannins produced by oaks are an excellent example of a quantitative defense: They do not poison a caterpillar after it eats them; they impede protein assimilation instead. This is a good defense because plant leaves contain very few proteins even in the best of circumstances, and leaf eaters cannot afford to lose any of these proteins because of tannins. Unless caterpillars have developed physiological adaptations to counter the effects of tannins, the more oak leaves a caterpillar eats, the fewer proteins it assimilates from those leaves.

In contrast to quantitative defenses, qualitative defensive chemicals are immediately toxic and typically require that caterpillars evolve specialized physiological adaptations, such as the acquisition of particular enzymes, before they can eat these compounds without dying. Monarchs, for example, can develop on milkweeds because the insects long ago evolved the ability to detoxify, store, and excrete cardiac glycosides, the poisonous compounds in milkweeds. The point is this: it is apparently much easier for insects to adapt to quantitative defenses such as the tannins in oak leaves than to adapt to qualitative defenses such as the cardiac glycosides in milkweeds, the cyanide in cherries, the nicotine in tobacco, and so on. It is likely that many characteristics of oaks—their large geographic range, their outsized apparency within ecosystems, the large size of the genus *Quercus*, and the reliance on more easily circumvented tannins as their primary defense—have all contributed to the large number of caterpillar species that rely on oaks for growth and reproduction.

I have four oak trees. In some years, they are very messy, raining acorns all over the yard and driveway; other years not so much. What causes this downpour in September?

The acorn downpour is called an oak mast. It is unusual behavior for trees, so ecologists have thought long and hard about why oaks reproduce so asynchronously. There are four hypotheses as to why oaks mast, but the two most favored are predator satiation and predator reduction. Predator satiation posits that masting is an adaptation against acorn predation. Acorns are such a valuable source of food for so many types of animals that if oaks predictably produced a moderate number of acorns each year, the squirrels and the deer, the mice and the jays, the ducks and the towhees, the acorn weevils and acorn moths, and all of the other creatures that rely on acorns to get them through winter would increase their population sizes to meet the available food supply. This would not be good news for oaks, because every year, large populations of acorn eaters would end up destroying nearly every available acorn, and oak reproductive success would plummet. But if oaks unpredictably and asynchronously produce many more acorns than there are acorn predators to eat them—that is, if they produce a mast—some acorns would escape the predatory scramble for acorns and germinate.

Predator reduction is another advantage for oaks that mast. During mast years, oaks produce unlimited food for acorn predators. This removes one of the biggest factors that limits population growth, so birds, squirrels, mice, deer, and other animals typically make more babies successfully during mast years than at other times. Unfortunately for all of these hungry new mouths, the year following a mast year is typically (but not always) a bust year for acorn production by oaks, and many acorn predators perish. This boom or bust approach to acorn production helps keep acorn predator numbers well below what would make oak reproduction iffy during most years.

Where did the word *mast* come from?

The Old English word *mæst* was the collective name for the fruit of beech, oak, chestnut, and other forest trees. It was derived from the Proto-Germanic word *masto*.

Do the rings in oak trees reflect mast years?

The short answer is yes. When oaks put a lot of energy into creating acorns, they usually do not grow much that year, resulting in a tree ring very close to the previous year's ring. Interpreting tree rings, however, can be tricky. Tree ring width varies for several reasons, including drought, defoliation, unusually cool summers, and so on. So you would need to have a complete understanding of these factors in the vicinity of the oak in question before you could pinpoint mast years from growth rings.

Is there a cycle of high/low amount of acorns according to weather extremes, drought/wet?

You are asking whether oak masts, the asynchronous production of acorns over time, is caused by the combination of certain weather events. The short answer is no. If the weather were constant every year, oaks would still mast in some years and produce few acorns in other years. But that is not to say that weather doesn't influence oak masts. Let's say, for example, that it's been three years since the last oak mast, and local oaks have accumulated enough energy to mast again. Flowering is prodigious, with lots of oak catkins making gobs of pollen. But just as the pollen matures and is ready to be released, a storm occurs with days of rain. Oaks are wind pollinated, so rain pretty much shuts down successful pollination. There will be no oak mast that year. The same could happen if a late freeze kills the oak catkins. So weather *can* play a role in oak masting, but it is not the dominant role.

Do acorn masts create large antlers in bucks?

They can. Acorns, an important food for deer, are high in protein, which encourages growth. We see the effect of mast production on antler size during the year following the mast—that is, last fall's mast crop will influence antler size this year. Antlers are highly influenced by nutritional quality, but genetics and age also play a part. Acorn mast is not the whole story.

Do all species of oaks mast in the same year?

No. Masts usually occur within an oak group (white oak group, red oak group, or live oak group). Occasionally, several species within a group will mast synchronously, but that is unusual and probably occurs by chance. Masts can be widespread, as was a large mast in 2019 that involved red oaks from Massachusetts to Georgia, and west to the Mississippi River; or they can be very local, sometimes involving only a single tree.

How do oak roots grow? Laterally or straight down in a giant tap root?

There is much talk about the "giant tap root" that an oak sends down and how that feature of oak growth makes them difficult to transplant. Young oaks do create a tap root in short order, but most root growth is lateral. The lateral roots and the root hairs they produce capture the nutrients required by the tree. Another urban legend is that roots stop their lateral growth at the dripline of the canopy. Root growth is a function of the size of the tree and the amount of moisture and nutrients in the soil. Root masses are smaller in wet soils because trees can easily acquire all the water they need with fewer roots. But in soils with typical moisture content, lateral root growth often extends two or three times the width of the dripline, which, in a large oak, can approach 300 feet from the trunk. With lateral root growth *that* extensive, the importance of a tap root is comparatively trivial!

Were oaks, periodical cicadas, and passenger pigeons ever connected ecologically?

Yes, indeed. Oaks and periodical cicadas are still connected ecologically. During the 2021 emergence of the 17-year cicada in Newark, Delaware, Josh Bernstein, an undergraduate at the University of Delaware, measured the oviposition (egg-laying) preferences of female cicadas by counting the number of flagged branches (branch tips that were killed after cicada eggs were inserted into the stems) on randomly selected street trees. Although the cicadas laid eggs in several tree species, the insects far and away preferred to oviposit in oaks. Cicadas develop underground as nymphs that suck fluids from the xylem of tree roots. It stands to reason that trees with the largest and longest lived root systems—oaks—would support the most cicadas.

Passenger pigeons, all three to five billion of them, depended on the masts of oaks (acorns) and beeches (beechnuts) to support their enormous populations. There is no doubt that the wholesale slaughter of pigeons by market hunters was the primary cause of the passenger pigeon's extinction, but the clearing of oak forests, particularly in Illinois, Indiana, and Michigan, where the pigeons nested, played a role as well.

Why do truffles prefer oaks?

Truffles are edible mycorrhizae that grow symbiotically with tree roots in both Europe and North America. They prefer oaks—particularly English oak, white oak, and live oak species—in France, but they do reasonably well on beech, Douglas fir, hickory, pecan, poplar, birch, and hazelnut too. Like other mycorrhizae, truffles grow only near the roots of these trees. When a truffle spore comes into contact with one of the tree's tiny rootlets, it produces microscopically thin filaments that wrap around the root. When this growth is complete, the root tip looks a lot like a small cotton swab and actually comprises both truffle and tree root. Mycorrhizal filaments are far better than bare tree rootlets at collecting nutrients from the surrounding soil, which they then share with the tree. But like all fungi, truffles are unable to synthesize sugars and other carbohydrates (they have no chlorophyll so they cannot photosynthesize), so they have to get their energy from their host tree. Thus, a productive symbiotic relationship exists: in exchange for some sugars and carbohydrates, truffles provide their host trees with the soil nutrients they need for rapid growth.

After this relationship is solidly established, typically by May, truffles produce a fruiting body (the part we eat) sexually, which grows throughout summer, and by September, it can weigh as much as a pound. These truffles ripen slowly into a black-and-white package of spores and by December are ready to disperse. This is where the famous truffle smell comes into play: truffles release an odor, largely a chemical called androstenone, that attracts truffle-eating mammals such as mice, wild boars in the old days, and pigs today. The spores pass through the digestive tract of whatever has eaten them to be deposited in feces under a new host tree, which, with any luck, will be an oak.

A 46-inch diameter, formidable bur oak is growing in my Minneapolis front yard. Is there a way to calculate its approximate age?

Short of coring it and counting the rings, not really. Oaks can grow far faster than most people realize when conditions (good soil, the right amount of

rain, favorable temperatures, no root disturbance) are right. If your yard provides a favorable environment for your oak, I'd guess that it is younger than
you might think. One indirect approach would be to find aerial photos of your
property (ask your local library). Often such photos exist back to the 1930s.
See if your tree shows up.

Which oaks produce the most insects?

No one has compared insect use among all of our 91 species of oaks. One of
my undergraduate students, Christian Stoltz, compared 16 species of oaks in
the mid-Atlantic States a few years ago by measuring the cumulative amount
of leaf area eaten by insects over the course of a summer. The idea was that
leaf area eaten by insects would correlate with the number of caterpillars,
katydids, and scarab beetles that had used each oak species as a source of food.
I can summarize his results this way: There was very little difference among
oak species. Members of the white oak group supported slightly more insects
than the red oak group, but the difference was minimal. The only oaks that
did not perform as well were water oaks and willow oaks, both of which are

Jays will fly more than a mile from the tree to bury an acorn for winter use, helping oaks
disperse faster than other trees.

southern trees that were planted outside of their normal range where Christian did his study. Thus, they were separated from at least some of the insect populations that typically use them for food.

How will oaks fare as the climate changes?

They'll do better than most tree species, because jays (Blue Jays, scrub-jays, Pinyon Jays, Steller's Jays, and others) pick up the best acorns and move and bury them farther away (up to a mile and a half from the parent tree) than seeds of other trees are dispersed. Those same jays select the largest and healthiest acorns (those that will provide the most winter food) to move and, in so doing, favor the oak genotypes that are best adapted to the wildly fluctuating climates we have inflicted upon Earth.

I have heard that oaks are promiscuous. Do acorns that fall from any of the oak species grow into that species?

By "promiscuous," I believe you mean different oak species can cross (hybridize) with each other. This is true within the various taxonomic divisions of the genus *Quercus*. For example, if pollen from a swamp white oak lands on a female flower of a white oak, the acorn that results will be a hybrid of those two species: when the tree grows, it will be its own unique plant, neither white oak nor swamp white oak. To be sure, the vast majority of the acorns produced by an individual oak will have been fertilized by a member of that oak's species. But when several oak species grow within pollen range of one another, hybridization does occur regularly.

Oak Management

I am fortunate to work at a school whose grounds are populated with glorious oak trees. Recently, the school has developed a zealous concern around addressing potential threats from some of the biggest trees. "Experts" have been hired to conduct radar mapping and resistograph data, which has led to the removal of two large valley oaks, and now we have received word of a third oak headed for the chopping block. Although I can well believe that, in some instances, a truly dangerous situation should be avoided by removing a sick or aged oak, I was struck by the following passage from your book *The Nature of Oaks*:

> We have been led to think that once there are hollow spaces created by rot within a tree trunk, that the tree must come down. Not so! Such "rot" is normal and does not affect the living cambium that lies just under the bark of your oak nor the functional strength of the trunk. Hollow trunks are just one feature of ancient oaks that makes them such valuable ecological additions to our landscape.

How can I help our school reach a better decision?

It does, indeed, sound like an overreaction by school officials. Try using this analogy with them: Pipes are hollow, yet still very strong. That is what an old oak becomes—a living pipe. The living part of any tree is the cambium, a thin layer of living cells on the circumference of the tree. All the rest is dead xylem tissues. Eventually, the dead xylem starts to rot away, but that doesn't weaken the tree. In fact, an oak with a hollow center can live for hundreds of years. It sounds like some lawyer is trying to create a risk-free world at the school. If officials are concerned with a particular old specimen, suggest that they put a fence around it to keep the kids away. This is a cheap and safe solution, and the world won't lose the ecosystem services provided by that great old tree.

Should I plant an oak where there is insufficient room for it to grow and reach maturity, knowing that the tree will eventually be cut down? Is five to ten years of ecological service worth it?

It is worth it. Oaks grow faster than you think and will reach a productive size in just a few years—and they support wildlife no matter their size. If your oak is cut down in 20 years, it will have helped your local ecosystem for 20 years, and that is certainly worthwhile.

I have an oak sapling about 5 feet high and a deer ate the top off. It is now growing with two branches at the top, and I think I should prune one side off so that only one grows straight up. When and how should I do this?

It is safe to prune your oak after the first hard frost.

Any ideas on how to stop city planners from clear-cutting oak groves for building?

Make it abundantly clear to the mayor that he or she will not be reelected if the city doesn't start employing ecologically sound land-use policies. If an elected official believes their job is threatened, they will act. Use social media to start a local campaign that demonstrates the public's support for maintaining oak groves. Let your voices be heard!

What is the most common critter that turns my oak's leaves into lace?

You are describing leaf skeletonization, which can be produced by many species of small caterpillars as they feed. My guess, though, is that you have a healthy population of pear slugs, the larvae of small tenthredinid sawflies that feed gregariously (in groups) without moving around much. They will skeletonize an entire leaf before moving to the next leaf. No worries, though, because a large oak produces 700,000 leaves and can easily spare a few for the pear slugs.

Pear slugs are gregarious sawfly larvae that skeletonize oak leaves.

How soon could I expect caterpillars to be present on an oak that's now only 3 inches tall?

Whether your oak is 3 inches tall or has a 3 inch trunk diameter, it can host caterpillars the first season it's in the ground. A few years ago, I came across a pin oak that had just put out its first few leaves, and standing on the stem was a caterpillar (a crocus geometer) stretching up to eat those leaves.

This little oak seedling provides a nice meal for a crocus geometer caterpillar, even though the tree is only 3 inches tall!

What can I do to preserve the large oaks in my garden in terms of gardening practices?

It may seem counterintuitive, but, in my view, the best thing you can do is nothing. Don't listen to the arborist who wants to prune them or treat them for "pests" such as bark beetles, or—heaven forbid—cut them down because they have some die-back or hollow spots, which are both normal for old oaks. Forego the urge to put in a backyard swimming pool or that unneeded addition to your house that will destroy half your oak's root systems. Do not allow heavy machinery—and that includes your riding mower—to cross their root systems for any reason. Instead, create no-go zones to help minimize soil compaction around those precious trees. And because going without oaks is a poor ecological option, I would get some young ones in the ground so that they will be ready to take over if the old trees give up sometime down the road.

What oak tree is safe to plant 15 feet from a septic drain field?

Small species such as dwarf chestnut oak would work. Or plant one of the many shrublike oaks if you live in the West.

My front yard is not big enough for a white oak or swamp white oak. Someone recommended a fastigiate cultivar of *Quercus robur*. Is this a good tree to plant or not?

The trouble with fastigiate forms of *Q. robur* is that these trees are cultivars of the English oak. They are not native oaks, and they are less able to support local caterpillars than a native oak species. Since making caterpillars is one of the most important contributions oaks make to local ecosystems, I would hate for you to have to compromise on that function for space reasons. Why not try a small native species such as a dwarf chestnut oak (*Q. prinoides*)?

When I type "water oaks and caterpillars" into Google, I see many results that include misinformation about caterpillars on oak trees, with information on how to get rid of them. It's really sad that so many misinformed people, even university professors, talk about how to get rid of

the insects that use oak trees. How are we going to progress if they are not informed of the facts?

Yes, the internet is full of university extension recommendations about how to get rid of insects. Any insect. The old mindset was that all insects were bad. It would be great to have such information removed from the internet, but many people still ask for advice about how to kill, kill, kill! This is changing, but far too slowly. The good news is that you recognized the role oaks play in producing the insects that feed birds and so many other creatures. You are not the only one who is enlightened, and the number of similarly enlightened folks grows every day.

Realizing that oaks add such value for wildlife, and keeping in mind what happened to chestnuts, elms, and ashes, do you recommend planting only oaks, or should at least a few additional tree species be planted too? Also, do you recommend that acorns to be planted be gathered from different trees versus many from one tree?

You bring up a good point: that a monoculture of any plant is vulnerable to introduced diseases such as chestnut blight and Dutch elm disease or invasive insects such as the emerald ash borer. I talk about oaks so much that people might think they are the only valuable trees in our landscapes. But that is not the case, even in places where oaks dominate naturally. If you have a large property, planting native cherries, willows, ashes, maples, birches, hickories, ironwoods, pines, firs, spruces, alders, hackberries, and other native trees along with your oaks (all biome dependent) is a wonderful idea because of the unique creatures those species could attract to your yard. Diversity begets stability, and in this rapidly changing world, we need as much stability as we can get.

Should we seek acorns from different trees to increase genetic diversity? That's not a bad idea, but is likely not necessary. Because oaks are wind pollinated, and oak pollen can blow more than 300 feet from the parent tree, there is a good chance that the acorn crop produced by one tree has, in fact, been created by the pollen from many other trees with built-in genetic diversity.

My wife asked me to take care of the bagworms that were attacking our oaks this morning. By that, she meant gather them up and burn them in the fire pit. As I complied, I got to thinking about your writing that oaks support lots of insects that, in turn, feed larger animals. What do you think? Should I have gathered/destroyed the worms by fire?

A far better approach would have been to leave the bagworms and let the parasitoids (such as ichneumon wasps) catch up with them and keep them under control. In my yard, I have to search to find a bagworm. Probably 50 percent of the bagworms you burned had natural enemies inside them that would have naturally controlled the population next year! No one was burning bagworms before we came along, yet the oaks were doing fine!

The common ichneumon wasp helps keep bagworm populations in check.

In the center of our front yard is a large oak tree. We have been dili-
gent stewards of our oak, with regular inspections and removal of dead
branches by certified arborists to ensure it's healthy and not a risk to
surrounding homes and passersby. This year, the arborist recommended
a tree-growth regulator (Cambistat) to reduce aboveground growth and
enable the tree to focus on belowground growth to strengthen the root
system. This chemical growth regulator is classified as a pesticide by the
USDA. Will it hurt acorn production and Lepidoptera?

I am sure no one has examined the impact of tree-growth regulators on the
insects that your oak supports. Cambistat is a systemic product, which means
it is translocated to every part of the tree. This also means insects cannot
eat your oak's leaves without ingesting this pesticide. Maybe it won't hurt
them, but I am skeptical. The real question is why would your oak need this
treatment? To slow its growth? I think your arborist is trying to sell you an
unnecessary, potentially harmful product to make some money. Perhaps I'm
jaded, but as I've said many times, beware of taking advice from someone
who will profit if you heed that advice.

**I am trying to plant as many oaks as I can in my soon-to-be-gone lawn.
What is the closest that I can space these very young trees in an area that
is begging for them?**

You will build a strong root matrix if their trunks are spaced 6 to 10 feet apart.
The trade-off is that before too long, their crowns will overlap, reducing the
total amount of sun each tree receives. Eventually, that will slow their growth.
But life is full of trade-offs! You are trading what would be a fast-growing,
isolated specimen tree for more stable tree groves.

**I have read your book on oaks and am fired up to help local citizens
plant oak trees in our community! Last year, my spouse and I bought
bare-root trees and gave them away at our local organic food store with
planting instructions. This year, we would like to order bare-root small
oak trees per your encouragement, as well as cages for tree protection.
We want to emphasize that owners must return each Earth Day for**

larger cages to help their trees grow safely and spaciously for the first several years. Can you please direct us to a place to buy progressively larger tree-protection cages? We want to have the materials handy to make successful oak growth as easy for folks as possible!

Well, pat yourself on the back! You are providing a great service. I recommend buying cage wire once rather than progressively. I bought galvanized wire (cheaper) and green-coated wire (prettier) at a big-box hardware store 22 years ago, and I am still using the cages I made then on new plantings. The wire mesh is about 5 feet tall. You can cut it to make a cage big enough to protect the oak until graduation (when you can remove the cage for good). You want the branches to be able to spread out right from the start. The most common mistake people make is to wrap a tree in a cage that is way too small. The branches grow faster than you think, and they will become constrained by a small cage before you know it. I support my oak cages with 5 foot sections of rebar, which can be purchased at any good hardware store. Deer are smart and will eventually figure out how to push your cage over if you don't stake it.

Should other healthy trees (such as walnut, cherry, and others) be cut down to allow for more or larger oak growth?

Oaks are great at performing many ecological roles, but they are not the only important trees in our landscapes. If you have a good diversity of native trees that includes oaks, I wouldn't cut down anything. But it depends on the density of trees you are talking about. If I had an oak that was surrounded closely by 15 black cherries, I would thin some of the cherries to give the oak some more light and space. I have a female American holly that was crowded by so many black cherry trees that it hardly flowered or fruited. After thinning a few of the closer cherries, the holly is now fruiting well. Black cherry trees grow much faster than hollies (or oaks, for that matter), so I will have to monitor the situation every year.

My sister and her husband just built a house near Pueblo, Colorado. For a housewarming gift, I plan to give them an oak tree. I learned that the Gambel oak is the only oak native to their area, but a few other oak

species will grow there. If it were my property in Missouri, I would plant several scrub oaks, but I want to gift them more of a specimen tree. What tree would you suggest and why?

My recommendation is to gift your sister one or more Gambel oaks so that they enjoy not just a tree, but the life those oaks will attract within the Gambel oak ecosystem. I understand your love of hardwoods, but if you try to re-create a Missouri landscape in the dry climate of Pueblo, you will lose the ecological value of that landscape. There is a reason Gambel oaks are the only oak species that exist naturally in most of Colorado—actually, lots of reasons, including the amount and timing of rainfall, winter temperature lows and summer highs, altitude, soil type, and more. If you plant a Missouri oak species in Colorado, some may survive, particularly with irrigation, but none will host the creatures the oak supports in Missouri. In other words, planting a non-native tree creates an artificial landscape that may be aesthetically pleasing but is ecologically foreign to everything around it. And that is what we have done ever since we started landscaping with ornamentals from elsewhere. Ecological function was not the goal, so we have created residential landscapes across the country with very little ecological function.

I am writing today from Portland, Oregon, a city well-endowed with trees from eastern North America, including red oaks. All of the eastern trees grow well, but none supports even a fraction of the food web that they support in the east. The leaves of Portland's red oaks are virtually untouched at the end of the season, because the red oak that is a keystone species in Pennsylvania contributes no energy to summer food webs in Oregon. It's beauty without function.

Should we be planting oak forests instead of conifer forests?

Only when oak forests are more appropriate than conifer forests. On the Eastern Coastal Plain, which runs from New Jersey south to Florida, millions of acres of mixed oak/pine forests have been converted to pine monocultures for wood and pulp production. If you are lucky enough to have the opportunity to restore one of these landscapes, you should definitely add oaks back into the mix. If you live in an area of the mid-Atlantic or Midwest where oaks have been logged below their pre-European density of about 50 percent of the forest, you can plant new oaks to increase their percentage. But don't try to force oaks into conifer forests in the northern part of the country, where they don't want to be.

When planting for wildlife, should you mix it up, with pin oaks, red oaks, white oaks, instead of several oaks of the same species?

You will want your oaks to be able to reproduce—that is, make acorns. Oaks cannot self-fertilize, so more than one individual of each species is needed to produce viable acorns. Oaks are wind pollinated, so if other individuals of the species are growing in your yard within pollen reach, your oaks can make acorns even if you have only one of each species. But if there are no other oaks nearby (within a few hundred yards), it is best to have fewer species and plant more than one individual of each.

Is it better not to plant oaks if we have to mow under trees because they are in a park?

In general, I would say it's *always* better to plant oaks, even if you can't create a layered landscape under the trees. But I question why, even in a park, the grass has to be mowed right up to the tree trunk? Couldn't a 10-foot area around the trunk remain unmowed? The tree would certainly benefit from that, and so would any caterpillars attempting to spend winter in the leaf litter on the ground below.

I am a botanist and certified arborist. I find myself wondering how much deadwood is acceptable to leave in the canopy of a large oak. If I were to look through an arborist's lens, I would prune regularly. But if look through an ecologist's lens, I can't help but notice the oaks in our yard are home to six of seven Missouri woodpecker species.

I'll remind you that I am not a certified arborist, but here is my take on your question. There are two types of pruning: deadwood pruning and live-wood pruning. There are no ecological benefits gained from pruning out deadwood that I know of; it is strictly an aesthetic pursuit and occasionally a safety one. But there are several downsides to pruning: Some 85 species of North American birds are tree-hole nesters, and they prefer to make their holes in deadwood. Moreover, many insects live in and help break down deadwood. In fact, the biomass of insects inside deadwood in an area is greater than the biomass of all the birds and mammals in that area combined. So deadwood supports a substantial amount of local biodiversity, and for that reason alone, it should preserved whenever possible.

Pruning live wood is a different story. Skilled tree pruning can prevent problems down the road that can shorten a tree's life. For example, if two equal-sized limbs originate from the same crotch, pruning down one of them can prevent crotch split later on. Arborists also prune to maintain symmetry. So pruning live wood is a judgment call. One word of warning: Never prune live wood during spring or summer. That can enable diseases such as oak wilt to infect the tree.

Do you bother to prune four-year-old oaks if they are in open spaces?

As a rule, I do not prune unless there is a good reason to do so (such as over-lapping branches). I am not an arborist, so pruning is not something I think about much. And I'm not interested in making my trees art pieces. Before we humans came along to "help out" our trees, the only pruners were the wind, ice storms, ovipositing cicadas (once every 17 years), and an occasional twig girdler. Somehow, most of the trees did well anyway, and many species lived hundreds of years longer than they do today.

What oak species are best for street tree use?

The answer depends on several factors: Is the area on the West Coast or Mid-west/East Coast? How much space is available for the planting? Usually, little space is available for street trees, which accounts in large part for their high mortality rate. Other factors include whether there are wires overhead (which there usually are), how much rain the area gets each year (urban trees are subject to dry conditions due to the lack of surrounding soil and heat island effects), and how deep the soil is. Shallow soils will send roots laterally and may cause uplift of nearby sidewalks or even the street pavement itself. Finally, soil acidity is a consideration. Urban soils are often quite basic because of the lime used in cement. So if all of these conditions occur, you should plant a small oak that does well in dry, basic soil. If you are in the

Midwest or East, that would be a dwarf chestnut oak. If you live in California, the valley oak is an option and is being used in an extensive effort to "re-oak" urban landscapes in Silicon Valley, even though this tree eventually grows quite large.

I have a lakeside property with 12 large red oak trees. The acorns from the trees near the driveway and house are problematic because they fall on our heads and ding cars. I want to neuter only the oak trees near the house/driveway. I understand there is a tiny window of time for doing this in spring. What do you recommend?

In theory, it is possible to use flora sprays that will prevent your oaks from forming acorns. But you're right: timing is everything. You have to hit the female flowers at just the right time to stop acorn production. And if you have big trees, it is difficult for homeowner-available spray equipment to reach the tops of the trees. You can hire a professional to take care of it, but that's expensive. I am lazy, and I'd take the easy road with this. I'd simply park my car someplace else for the two weeks of the year that red oak acorns fall. Red oak acorns take 18 months to develop, so a tree that dropped a lot of acorns one year will drop very few the next year.

My question is about non-native oaks, particularly the sawtooth oak and English oak. Several suburbs and neighborhoods around Columbus, Ohio, have planted these non-native species as street trees, and the parks supervisor of my suburb is keen on planting them in our community. I have read that the sawtooth oak reaches reproductive age quickly and has the capacity to be invasive, and the English oak takes a very long time to reach reproductive maturity and is unlikely to become invasive. Are either of these species beneficial to our native moths and butterflies?

You're right to question the use of sawtooth and English oaks in Ohio. Sawtooth oaks are from Asia, they're invasive, and their acorns, though plentiful, appear to be too bitter for many acorn eaters. Their leaves are rarely touched by native insects, although Japanese beetles like them (if you want more Japanese beetles). I have never compared English oak to native oaks, but my question is, with 91 North American oak species, why would we *ever* need to plant exotic oaks, which are sure to provide less for wildlife? Remember that oaks drive food webs by producing caterpillars, and native oaks do that much better than non-native species.

I have a relatively small yard. The squirrels have gifted me with bur oak saplings. I have moved a few to locations where I want big trees. One oak is now almost as tall as my house and the birds love it. The other oak volunteers in my yard are 4 or 5 feet tall or smaller. I am hesitant to pull any of them, hoping they will decide which trees will thrive. Is there a limit to the number of oaks my small yard can support, and will the trees make that decision for me?

I would vote for managing your little woodlot to favor the most productive plants. As for whether to let your young oaks duke it out with each other, it does sound like you have more trees than you have room for. If this were my yard, I would leave the small oaks with hope that the deer in my yard would eat them instead of the larger oaks.

You helped me with advice regarding the planting of three white oaks on our property last fall. They are planted about 10 feet apart in what was regularly mowed yard. I know that you believe that the surrounding groundcover is important. What can I plant in the ground below the trees that might be best?

I would create a bed under the trees by laying down a layer of leaf litter. Even if you planted nothing in the litter, it would function well in helping caterpillars complete their development and in returning appropriate nutrients to your oaks. But you could add color right away in the form of wood poppies, which spread quickly and have greenery throughout most of summer, even after the yellow flowers stop blooming. Wild blue phlox creates a beautiful groundcover when in bloom in spring, but many other groundcovers would work as well. Try foamflowers, wild ginger, Virginia creeper, mayapples, ferns, or a mixture of these. The species that favor the conditions the most will do the best and give you the cover you are seeking.

How can naturally growing bur oaks best be transplanted, and how far do their roots extend? We have lots of 3-foot baby oaks under a

140-year-old bur oak on a pure sand bank along the river. We want to keep the genetics going.

Your best bet is to start with next year's crop of acorns from your tree. A 3-foot seedling growing in shade could easily be more than 20 years old and have a substantial root system. It's not impossible to transplant the seedlings, but it is impossible to transplant them without a lot of root pruning. That will set them back so much that starting from an acorn will result in a healthier tree in the long run. If you want to try anyway, move them as soon as the ground thaws in late February.

The nursery industry heavily promotes hybrid oaks, claiming faster growth and quicker acorn production. But is this true? I looked for research and couldn't find any, so I started my own. So far, I see no difference. Do you know of any differences between pure strains and hybrids when it comes to supporting insects?

Oaks do hybridize naturally among species within a taxonomic section—that is, members of the white oak group can hybridize with each other, members of the red oak group can cross with each other, and so on. If those crosses make more acorns than straight species, I haven't heard about it. I would avoid crosses between North American oaks and non-native oaks like English oak or Asian sawtooth oak. The foliage of such crosses will not support oak-dependent caterpillars nearly as well as the leaves of straight species.

How can we encourage getting Georgia oak into commercial production?

Georgia oak is one of the few species of oaks that does not reach enormous size at maturity, which makes it ideal for small landscapes. Unfortunately, it is one of our oak species that is endangered, largely because it has such a small natural distribution: it is found only on dry granite and sandstone outcrops of sloping hills. But Georgia oak tolerates a variety of soils, as long as they are well drained and in full sun. Surprisingly, it has been in the nursery trade in the past, and I hope it will play a more prominent role in the future. But someone has to go to the trouble of creating a reproducing stock. I suggest that you ask for it at every nursery you visit. If we can convince nursery owners and growers that there is a market for this tree, someone will take action.

During your talks, you discuss installing plants under trees. I planted a native woodland shade garden about 20 feet from my pin oak. I also planted native ferns and wild ginger under the oak with 1–2 inches of mulch. I'm reading that maybe I shouldn't plant so close to the trunk—that the oaks don't like having anything planted by their roots. Is this correct?

Planting under oaks is not problematic as long as you do it in a natural way. There's not much soil close to the trunk, so a layer of leaf litter to catch moisture during winter is the best option. Planting shallow-rooted species such as ferns a few feet from the trunk is fine as long as you don't dig so deep that you damage the oak's roots. And don't plant anything under an oak that will require irrigation during summer; overwatering is a problem for oaks in the drier areas of the country. Some people plant flowers under oaks and then water them all summer long, but oaks don't tolerate being watered in the dry season, and doing so can cause fungal attacks. In short, it's not that oaks don't tolerate plants near their roots; it's that they don't tolerate unnatural disturbance, fertilization, or irrigation near their roots. But 20 feet from the trunk is not too close.

In *The Nature of Oaks*, you mention growing a "grove of trees," which makes complete sense to me. But do all of those trees necessarily need to be of the same species, or can I mix it up? If they are all oak trees, I should space them about 10 feet apart, correct? If they are not all oaks, then how far should I space them?

All the trees in your grove do not need to be of the same species to interlock their roots, just as all of the trees in a forest are not of the same species. The spacing can vary, but small trees should be planted closer to each other than trees that will eventually become large. Ten-foot centers for larger growing tree species is a good spacing.

Can you please explain more about creating a soft landing zone under oak trees?

You are referring to the fact that more than 90 percent of the caterpillars that develop on oaks drop from the tree after they have finished growing and then either pupate underground or spin a cocoon in the leaf litter under the tree.

We are still studying the conditions that best favor the survival of species that pupate underground, but the most important factor seems to be soil compaction; it is very difficult for these soft little caterpillars to push their way into soil that is compacted. When they can't find loose soil under the tree, they just start walking until they do find it—or until they get picked off by a bird or squished by a foot, lawnmower, or car. Species that spin cocoons in leaf litter need—you guessed it—leaf litter. The goal, then, becomes creating beds under our oaks—under all of our trees for that matter—that provide two things: uncompacted soil and leaf litter. You can achieve soil conditions that facilitate caterpillar pupation by reducing traffic under your trees, by adding organic matter under the trees in the form of the leaves that fall from those trees each year, and by planting groundcovers with penetrating roots. These could include any of the spring ephemerals such as mayapples, goldenseal, ferns, foamflowers, Jack-in-the pulpit, Jacob's ladder, wood poppies, phlox, native pachysandra, and dozens more options appropriate for your region.

Densely planted groundcovers under your trees provide excellent pupation sites for caterpillars.

How can I control the boxelders growing under my oaks?

A few boxelders growing under your oaks is not a bad thing. It creates a layered landscape, just as you would find in a healthy forest. Ideally, our landscapes would contain all of the vegetative horizons found in forests: the canopy, subcanopy, understory, shrub, and groundcover layers. If you have too many boxelders, get out your loppers and cut a few down at the ground level.

Are there any small straight-species oak trees that could be planted in a very small yard, that are also long-lived? I have been told that small oak species are pioneer species and would die off in 20 years or so.

Blackjack oaks can live for more than 200 years. The dwarf chestnut oaks in my yard are 23 years old and look great. Some of the oldest living organisms on the planet are ground-hugging oaks in California. The palmer oak, for example, can live to be 13,000 years old!

When is the best time to relocate a young oak tree that's 7 feet tall?

That is a major transplant event that will involve quite a bit of root pruning. Early spring—say, March—is probably the best time to attempt something like this. The leaves will not have been produced yet, which is a good thing, because their size and number will somewhat adjust to the loss of root biomass resulting from the transplant.

Please discuss the best way to start bur and white oaks from acorns, how to overwinter them (in the upper Midwest), and when to plant them into oak savanna/woodland restorations.

Congratulations on deciding to start your oaks from acorns. Acorns are easy, free, and plentiful, and they will grow into healthier trees than transplanted established trees. The only thing they don't offer is instant gratification. But if you stick with your decision, here are a few hints to help you out. First, know which species you are planting. White oak acorns germinate in fall, just days after dropping from the parent tree. These should be planted about a half-inch below ground soon after you collect them—pretend you are a squirrel or a Blue Jay storing your nuts for winter. Your acorns will send down

When planting an acorn, don't plant it with the root end facing up!

their tap roots in fall but will not pop above ground as epicotyls until spring. Also, take note of which end of the acorn the root will emerge from (it's the more pointed end), and be sure to plant that end pointing down in the soil. If you aren't sure which end is which, plant the acorn on its side. The root will emerge and head down with no trouble.

In contrast, bur oaks do not germinate until spring. But don't wait until spring to collect bur oak acorns; usually the only ones you'll find in spring will be inferior or damaged acorns already rejected by myriad acorn eaters over winter. Either plant them as soon as they fall or store them in a sealed plastic bag filled with damp peat moss in the refrigerator until mid-March. This will protect them from hungry rodents during the long winter months.

This brings us to an important decision regarding planting acorns. If you plant them directly into the ground at their final resting site, chances are good they will be found and destroyed by some critter before they germinate. For this reason, you should consider planting acorns in deep flowerpots in a mixture of local soil and well-draining potting mix. Protect your pots from marauding mice by storing them in a place that is exposed to some of winter's cold, but not the full might of a polar vortex. Remember that the potted acorns are above ground and won't receive any moderating warmth from the unfrozen soil beneath them, as they would if they were planted directly in the ground.

Another winter challenge is desiccation; if you store your potted acorn where it will not be exposed to seasonal rain or snow, the soil may dry out. You can avoid this problem by lightly watering it about once a month. In spring, wait until your seedling has fully expanded its first true leaves before transplanting it into the ground. But don't wait too long; oaks grown in pots quickly become rootbound, a condition that often kills the tree even months or years after transplanting. This all may sound pretty complicated, but it's really not that difficult, and in my view, it's well worth the thousands of dollars of savings that would otherwise go toward buying a large tree.

I live in central New York, where the soil is alkaline. We moved from Massachusetts, where we had acidic soil, and oaks grew wonderfully there. I have blueberry bushes, but they are small and I acidify the soil with a wash every year. How many years would I need to do this to insure that oaks would grow on my property?

I would *never* try to acidify the soil. That is fighting nature, and it's often a losing battle. A much better idea is to go with the conditions you have. You can grow oaks in your soil by choosing the right plants. Several oak species will grow well in your alkaline soils, including chinkapin oak, scarlet oak, dwarf chestnut oak, and bur oak.

Do oaks need to be watered during extreme droughts?

This is a tough question. Certainly a young, transplanted oak that has not yet rebuilt its root system would need supplemental water, but an established oak tree with a full complement of roots with a spread of two or three times the width of the canopy? Decades ago, during a severe drought in the mountains of Virginia, I witnessed mature white oaks wilting. The same thing happened in parts of New England last year. I'm sure those wilting oaks would have loved a good drink of water, but when the rains finally came, the trees bounced back and still stand tall today. And as you move west, oaks are adapted to low rainfall, especially in the summer months, and they can actually be killed by supplemental watering at the wrong time of year. I suggest you use your own judgement on whether or not to water an oak during a drought. If you do water, though, do it cautiously.

Our co-housing community will be planting a good number of trees this fall, and we want to plant large trees. What is the largest size of oak trees for planting that you'd recommend?

I recommend you plant trees that are 2–3 feet tall! I know that's not what you want to hear, but I am totally against spending lots of money on a large tree that will grow very slowly for years while it tries to rebuild the roots it lost during transplant. You can get a healthier, faster growing tree for far less money if you start small. Delayed gratification pays off big time when it comes to planting oaks.

Large trees cannot be transplanted or grown in pots without reducing the root mass necessary to support the tree, and many die after planting, as happened to these large trees planted in Newark, Delaware.

My family has kept a large oak woods growing for harvest in southern Iowa since 1854, with the help of district foresters who mark them for us "shelterbelt style" in exchange for us keeping the woods healthy for wildlife and in the American Tree Farm System. After the last harvest, a few large oak trees unexpectedly died, possibly resulting from the stress of the others being removed, and there had been droughts for three of the previous five years. With warming weather, oak wilt, and invasive plants, how hard is it going to be to keep these large family-owned hardwood forests going and at least somewhat profitable so woods aren't converted to pastures or crop fields?

Well, it won't be hard to keep your woods going, as you say, even if oak diseases attack the majority of your mature oaks. Some trees will be resistant and will produce the acorns that will slowly replace the susceptible oaks in your woods with resistant genotypes if they're allowed to grow. It's the "somewhat profitable" part of your woodlot that is the challenge. It wouldn't be a challenge at all if you were paid for the ecosystem services your woodlot produces for you and your community—for the oxygen, clean water, flood control, and carbon capture it provides year in and year out; for the abundance and diversity of animals and pollinators it houses and feeds; for the moisture it adds to the air each summer day; and for the moderating effect it has on severe weather. The wood it produces is actually the least valuable of its many contributions, even though it is the only thing we are willing to pay for. How wonderful it would be if you got a tax break or USDA Conservation Reserve Program (CRP) supplement large enough to keep you from converting yet more of Iowa to cropland (85 percent of Iowa is now in cropland).

I believe that someday we will enact policies that will enable you to keep a woodlot a woodlot, without the need for any wood harvesting. Maybe establish a "pay-per-use fee" (10 dollars a year?) for everyone who lives in Iowa City, Des Moines, Cedar Rapids, and all the other urban areas in Iowa—some three million people who use the ecosystem services produced by woodlots. That would provide 30 million dollars annually to help you and others like you save your woodlots. Just 10 dollars per year! That's a couple of trips to a coffee shop! It is past time that we took care of the natural systems that take care of us. Pass it on!

I want to move random oak seedlings that have come up in my gardens/ lawn this summer. Would moving bare-root trees this fall or this spring be the best idea? I don't want to pot them and then replant. Also, considering I have jumping worms, I would brush and wash off the roots before replanting in a restoration site.

Yes, that should work. As you dig up the oaks, try to save as many roots as you can (especially the root hairs). Gently hose the soil off, and then transplant as you would with any bare-root plant. I would do this in early spring as soon as the ground thaws. Jumping worms overwinter as tiny egg sacs, so your objective is to make sure there are no egg sacs left on the root mass. Jumping worm egg sacs are round, black, and only 2 or 3 millimeters wide; they look very much like a seed or a smooth clump of soil. It is nearly impossible to detect them all with any confidence.

Given the importance of flowering time, leafing out time, and initiation of dormancy, how important is local ecotype sourcing of oak trees?

Planting local ecotypes is important for exactly the reasons you state: local genotypes are adapted to provenance—the environmental characteristics of your latitude, longitude, altitude, and soil type. Matching provenance is particularly important for wind-pollinated plants. If you plant an oak, you will want it to become a functional member of your local oak community, and that means it must shed its pollen and expose its female flowers exactly when surrounding oaks do, or it will never produce acorns. If you plant an oak that germinated from a seed 500 miles south of where you live, it will be genetically programmed to welcome spring when the oaks 500 miles south welcome spring. Asynchronous phenology!

Oak Diseases/Pests

What are the differences between sudden oak death, bacterial leaf scorch, and oak wilt? Are they the same?

They are distinct diseases that we humans have inflicted on North American oaks in recent decades. Sudden oak death (SOD) was introduced to California from Asia on nursery stock in the 1990s and is a terrible scourge that has already killed at least a million oak trees. This deadly fungus is particularly hard on coast live oak and tanoak, an oak relative, and has devastated coastal forests in California and Oregon. The disease also infects rhododendrons, camellias, and other common horticultural plants that have served as "typhoid Marys" as nursery stock is moved around the country. In April 2004, SOD was found in 14 nurseries in Georgia, and it has since been detected in Florida and in nurseries in more than 20 states. The nursery industry is poised to degrade ecosystem function seriously throughout eastern forests, just as it did by introducing the chestnut blight.

There is some good news, though. A 2020 study by A. O. Conrad suggests there is, indeed, resistance in coast live oak (27 percent of the trees tested in this study) and it is heritable. And this seems to be the general pattern for most plant diseases. Even when most of the plants in a population are highly susceptible to a pathogen and die quickly, some proportion of the population, albeit often quite small, will have at least some resistance to it. That is the genetic stock that will expand over time to create populations that are mostly resistant. This is also why I am so opposed to advice from arborists and foresters to stop planting tree species that are currently under attack by diseases, such as oaks, beeches, ashes, chestnuts, and American elms. Rather than not planting these trees, we need to plant more of them than ever. Yes, many will die, but the resistant genotypes will not die and will serve to replace the susceptible genotypes that are lost.

Bacterial leaf scorch (BLS) is a systemic disease caused by the xylem-plugging bacterium *Xylella fastidiosa*. It is most commonly seen in pin, red, shingle, bur, and white oaks, but it can also affect elms, oaks, sycamores, mulberry, sweetgums, sugar maples, and red maples. Xylem-feeding leafhoppers and spittlebugs spread the bacterium from tree to tree. Transmission between trees through root grafts has also been reported. There is no cure for this disease, and it is often fatal. Where did it come from? No one seems to know, but BLS is described as a "new disease," which to me clearly means it was introduced on a plant that was brought to North America from another continent.

Oak wilt is a highly transmissible disease caused by fungal pathogen *Bretziella fagacearum*, which attacks the vascular systems of oak trees. Like leaf scorch, it causes leaves to brown and wilt. It is a particularly serious problem for members of the red oak group. In red oaks, leaves in the upper crown are the first to become discolored and wilt, followed by leaves lower in the crown. Unlike leaf scorch, oak wilt often causes browning of half or more of the leaf area at once, including veins. Infected leaves soon fall, defoliating the tree in just a few weeks. In white oaks, the disease proceeds more slowly, often beginning with the death of a single branch, followed by other branches over the years. Oak wilt is highly transmissible, both through interconnections between the roots of neighboring oak trees (root grafts) and by several species of bark beetles that carry fungal spores from one tree to another. These beetles tunnel into trees through fresh wounds, which they are able to find in minutes. This is why you should not prune your oaks when beetles are active, from April until at least mid-July, and ideally not until the first hard freeze. Oak wilt was first detected in Wisconsin in 1944 but has spread rapidly since then, a clear indication that it is a disease that was brought here from another continent.

Does the two-lined chestnut borer pose a threat to oaks?

The two-lined chestnut borer is a native species of metallic wood-boring beetle that develops within oaks that have been stressed by drought, storm damage, or disease. They are never the primary cause of oak decline. Its name comes from that fact that it developed in American chestnuts before the chestnut blight all but removed them from eastern forests. With the droughts, storms, and diseases attacking our oaks of late, the beetles are having a heyday. But they are a secondary issue. The best way to keep our oaks healthy and limit populations of two-lined chestnut borers is to favor oak genotypes that are resistant to bacterial leaf scorch, sudden oak death syndrome, and oak wilt. Treating for two-lined borer will not save your oaks if they are sick from disease, but it will enrich your tree care company's profits. Such treatments are not only a waste of money, but they poison your oaks for the other creatures that use them, seriously reducing the trees' ecological value.

If there is anything good about the spate of oak deaths that is occurring across the country, it's that we now have the opportunity to add ecologically valuable snags to many landscapes. Snags (dead, standing trees) used to be a feature of every wooded landscape, but for aesthetic and sometimes safety reasons, we typically cut down trees as soon as they die. In fact, we typically cut down trees at the first sign of decline. Dead and dying oaks are valuable additions to your landscape because they provide niches for many species—probably not more species than living oaks, but a lot. For example, 85 species of North American birds breed in tree cavities, and dead oaks create wonderful opportunities for many of them, but only if we leave the trees standing. A dead oak can provide valuable wildlife resources for decades.

I planted some oak trees as seedlings 10 years ago and they are still no more than 2 feet tall. What should I do to enhance their growth? I know oak trees are slow growing, but this seems exceedingly slow.

It sounds to me like your oaks are growing in the shade. Oaks will bide their time in shady conditions, just waiting for a tree fall to create a light gap—an area of sun large enough to provide the trees with several hours of sun each day. That's all they need to stimulate rapid growth.

My arborist says my oak has lace bugs and he wants to treat the tree with systemic insecticides. Should I let him, or will that hurt the tree's ability to support other insects?

The only thing your tree will support after treatment with a systemic is your arborist. He almost certainly will use a neonicotinoid, an insecticide that's 7000 times more toxic to insects than DDT, and it will render your tree a dead zone for three or four years. The oak, which could be a factory that produces the caterpillars needed by birds in your yard to feed their young, won't produce anything. I also worry about the impact on the birds that eat the tree's acorns. I don't think it has been studied, but systemic insecticides are usually transported to all parts of the plant, including its fruits, seeds, and nuts. But here's the thing: Lace bugs won't hurt your oak at all. All self-respecting oaks in the white oak group support populations of oak lace bugs with no ill effects. Your arborist is purposely or inadvertently misleading you so that he can make a buck.

I am determined to create a yard that feeds wildlife even though it hasn't proven to be a realistic goal. Between the groundhogs, rabbits, drought, deer, invasive insects, and neighbors, it's been hard. What large tree can I plant in dry, sandy, alkaline soil and full sun that won't succumb to disease or insects in 10 years? Hackberry?

Two oaks would work for you: chinkapin oak and bur oak. Both do well in alkaline soil. Start with small trees so they can develop a full complement of roots. That will give them the best chance against oak diseases and dry soil.

My property is next to a farm, and my oaks are exposed to pesticide drift/ aerosols. The leaves are very curled this year. Any advice?

Your oaks are the unfortunate victims of oak tatters, an affliction caused by vaporized chloroacetamide herbicides, primarily acetochlor. Dr. Jesse Randall of Michigan State University has empirically demonstrated that when young leaves of white oaks, hackberry, and a few other trees are exposed to metolachlor (dual herbicide), acetochlor (harness herbicide), and dimethenamid-P (outlook herbicide), the interveinal tissues on the leaves disintegrate, making the leaves look tattered. Seedlings are particularly vulnerable. Years ago, it was thought that tattered leaves were caused by cold snaps in spring, but the evidence suggesting herbicides are to blame is much stronger. What is frustrating is that there would be no volatilization (vaporization) of these products if they were encapsulated to prevent this—that is, we know how to prevent yet another assault on oaks by changing these products' form, but manufactures so far have refused to take this extra step. Why? Because it costs a bit more! But think about it: if enough herbicide is vaporized in the air to melt oak leaves, what is it doing to our lungs? I suggest you write to your legislators and let them know that this is an issue you care about, and that you will be voting with it in mind.

Connecticut had a bad summer with the spongy moth caterpillar last year. I am the lucky owner of one beautiful, very large white oak that was nearly defoliated last summer. I am worried that if we have a summer of spongy caterpillars at all close to last year's numbers, it could be fatal for this old oak, and I've been weighing the ecological pros and cons of using an injectable abamectin pesticide, which can be applied by an

Oak tatters is a disorder caused when pre-emergent herbicides volatilize and drift out of the agricultural field on which they were applied.

arborist. I know the pesticide is indiscriminant and will kill all oak-dependent insect populations. If you were me, would you use it?

If your tree is loaded with egg masses, you will likely have a good deal of defoliation, even if the fungus that controls spongy moth breaks out this spring. So, yes, I would treat the tree, but not with a systemic pesticide like abamectin. Systemics tend to last a long time, years in many cases, and you may not have any spongy moths in future years. Abamectin is also highly toxic to bees that collect oak pollen, even though oaks are wind pollinated. I would use a standard spray of *Bacillus thuringiensis* (BT) instead. Yes, it will kill all the caterpillars on your oak, but only this year.

Does road salt hurt oaks?

Very few tree species like a lot of salt, and road salt treatments can leave tons of salt where there should be none. But if we were to compare which oak species can best handle this physiological insult, we would find that members of the red oak group are slightly more tolerant.

With the spread of exotic diseases such as oak wilt and sudden oak death, is it, in your opinion, just a matter of decades before our eastern oak species end up like our native ash species? Just the thought of something like sudden oak death indiscriminately killing every unprotected oak tree on the East Coast is so horrifying.

Heavens! We cannot let our oaks fall victim to our thoughtless introduction of diseases from around the world. They are simply too important ecologically to lose them from our forests or our managed landscapes. Fortunately, all of our oak populations are showing various levels of resistance to these oak diseases. We will lose many trees, but the acorns of the resistant ones will be dispersed by jays and other animals. Perhaps a bigger concern is the overabundance of browsing deer that are preventing oak seedling recruitment (the addition of young plants to the population) into our forests.

My white oak just died from what an arborist told me was an ambrosia beetle infestation. A red oak next to it is also dying and also shows signs of ambrosia beetle activity. Is there anything I can do to save it?

My guess is that your oaks are dying for another reason, probably from one of the oak diseases that have been brought into this country, and the ambrosia beetles are just taking advantage of the weakened trees. It's not the ambrosia beetles that are killing your tree, so treating for the beetles won't help the tree. Unfortunately, our ability to save trees that are infected with oak diseases is poor. People often invest big dollars in tree care, and then the oak dies anyway. The key is to favor resistant genotypes that do not readily succumb to diseases.

Red oaks are susceptible to oak wilt. Normally, trees aren't treated when you prune. If you have storm damage during the wilt months with a red oak, would it make sense to treat it? If so, what product should be used?

Oak wilt spores infect wounded trees most readily between April and July, and you should never prune in those months. But to be safe, you should avoid pruning during the entire summer, and I suggest you wait until after the first hard frost to prune. If your tree is wounded during the summer months, it is worth the trouble to treat the wound. Coat the wound with a thick layer of petroleum jelly as soon as you can. That keeps the disease spores and insects that move them around out of the wounded area until it heals over.

I have a pin oak in my front yard. I noticed it has boring insects that are leaving scars in the bark. Thus far, they have not killed the tree. In the interest of preserving more native insect life, do I treat the tree or just let nature take its course and if the tree dies, it dies?

My guess is that the holes you see in your tree are being drilled by the Yellow-bellied Sapsucker, and not a boring insect. The sapsucker drills shallow holes when sap is running and then drinks the sap. These holes will not hurt your tree. In fact, they are a sign that your tree is supporting a sapsucker over the winter months, which is something to celebrate and enjoy!

What causes the big burls on oak trunks?

Burls on oaks (and other tree species) have many causes, including fungal and bacterial infections, viruses, and bark injury. A burl is essentially a tree's

Burls on oaks are caused by infection or injury to the bark, and they can be huge.

wound response. They can be quite large and can be used for fashioning certain types of furniture. However, if you cut a burl off of a living tree, the tree will most likely die.

Oaks are dying right and left. As we cut them down, should we have the logs hauled away to prevent infection of other trees? Or can we use the logs on site?

Please use the logs on site. Disease spores float for miles on the wind, so hauling the logs away won't help your other trees. The best ecological use is to leave them on site as "coarse woody debris." Species-rich communities live in and depend on dead logs and branches on the ground. This is especially true for beetles: entire families of beetles specialize on fallen logs for development, including the iconic stag beetles, bess beetles, fire-colored beetles, many click beetles, and our most awesome rhinoceros beetles. Coarse woody debris also serves as nesting sites for many species of native bees and provides overwintering shelter for butterflies such as the mourning cloak, comma butterfly, and question mark butterfly. And, from the perspective of Pileated Woodpeckers, our largest and most spectacular woodpeckers, dead logs are essential, for they house colonies of carpenter ants, the sole source of nourishment for their nestlings. A Pileated Woodpecker is, in essence, nothing more than thousands of carpenter ants dressed up in pretty black-and-red feathers!

The stag beetle (left) and rhinoceros beetle (right) are just two of the many creatures that depend on fallen logs.

What is the greatest threat to oaks?

If you have read this far, you will know that imported diseases such as oak wilt, sudden oak death syndrome, and bacterial leaf scorch are hammering oaks nationwide. These diseases are certainly enormous problems, but, as it does with most other forms of biodiversity, human development picks away at oak stands every day of the year. Less obvious is the impact of deer on oak populations. When deer are too numerous, as they are in most parts of the country, they devour and decimate oak seedlings, seriously limiting recruitment of new individuals into woodlands. It is hard to say which of these threats is the greatest; they all threaten the future of our oak forests.

Symptoms of bacterial leaf scorch include half-leaf browning in midsummer.

I know that oak wilt affects red oaks predominantly, but how does it affect the white oak group? Can we continue to plant trees in that group?

Yes, please do plant white oaks and other members of the white oak group (post oaks, chestnut oaks, bur oaks, chinkapin oaks, and others). They have a great deal of genetic resistance against this particular disease.

What are the galls on my chinkapin oak?

There are hundreds of possibilities. Galls are created when tiny cynipid wasps lay eggs in the meristematic tissue of oak buds, leaves, and stems. Along with the eggs, the wasp injects plant hormones that partially control the growth of the undifferentiated cells in the tree's tissues. Oaks also have a say in how those cells grow, the result being a species-specific growth we call a gall. Every cynipid species creates a gall uniquely shaped for its use. The larval gall wasp develops within that gall. Worldwide, more than 1000 species of cynipid galls are associated with oaks, 300 of which live in North America. A single oak tree can host up to 70 species of galls.

Tiny wasps in the family Cynipidae create the galls that are so common on oak trees.

What is the effect of global warming on oak trees, and what can we do about it?

The impact of climate change on oaks will depend a lot on geographic location. Oaks in California are being clobbered by megadroughts and the resulting forest fires. Oaks in Massachusetts were recently impacted by a drought that lasted several years and triggered a spongy moth outbreak. The worst derecho (straight-line wind) on record leveled many thousands of oaks in eastern Iowa in 2019. I could go on, but you get the idea. Extreme weather events are hard on all life, including oaks.

What can we do about climate change? As individuals, the most effective thing we can do is support legislation (and legislators) that facilitates the transition from fossil fuels to clean energy. I can share one bit of good news, though. I recently returned from a trip to southern California, where a large crown fire had roared through a canyon a few years earlier. Every tree was charred to its very top, and all looked quite dead. Yet despite the severity of the fire, each oak sported new growth coming up from its roots! They were not dead after all!

I have a very large pin oak that is infested with horned oak galls in large quantities. The oak is shedding twigs of leaves *all* the time. Several on my street are also infested, and one neighbor had their tree removed recently—I really don't want to do that—it's a beautiful tree! Is there anything I can do to rid it of the galls? Will it have to be cut down?

Parts of the Midwest, particularly the Saint Louis area, are reporting big outbreaks of horned gall on pin oaks. Since gall wasps host more parasitoid species (tiny wasps that kill gall makers) than any other insect group, I have wondered why horned galls are not being controlled by their parasitoids in these areas. They certainly are at our house. Out of our 10 pin oaks, I counted a total of three horned galls. Could it be that mosquito fogging is wiping out the parasitoids in Saint Louis? If so, that would be another good reason to ban mosquito fogging, which, by the way, does not control mosquitoes! But to answer your question, horned galls are built from woody material and are essentially part of the tree. Even after the gall makers have left, the gall remains for the life of the tree (or its branch). There is no effective way to prevent your tree from being attacked. The best way is to let nature control them, but you can't kill natural enemies by spraying and then expecting good control.

The oaks on my property go from new, whole, unblemished leaves in spring to badly damaged leaves by late summer. If trees communicate pest predation to each other so they can mount chemical defenses, is it the vast insect onslaught that creates the leaf damage or something else about an oak's ability to protect itself?

Plant defenses are not perfect, and I'm glad they aren't. If they were, there would be no animals on Earth, and that includes us. Most plant defenses are best at protecting against generalist insects that do not have all of the adaptations required to circumvent or disarm them fully. Specialists, in contrast, are good at disarming their host plants and sometimes use host-defensive chemistry to find their hosts among a sea of inappropriate species.

Oaks depend on leaf toughness, and tannins are their primary defense. Tannins are apparently easy to evolve, because lots of plant lineages employ them. Their general mode of action is to make protein absorption difficult in the insects that eat them. By regulating their gut pH, however, many insects have found a way to reduce the effectiveness of tannins as a defense. As you say, lots of insect herbivores have adapted to oak defenses, which is why oaks are the best plant genus in terms of passing on energy to food webs. The damaged leaves on your oak trees are a testament to your oak's productivity.

Should basal shoots or suckers on a young, damaged oak be pruned? Are they hurting the tree (draining resources from the canopy) or helping the tree (supplying more resources to the roots to rebuild the health of the tree)?

I vote for your second explanation. When a tree is damaged, it attempts to repair itself. Deciduous trees are far better at this than are conifers, and if the injury is severe enough, they will send up shoots from the base of the tree. This is why cutting off a tree near the ground usually produces a copse. So many shoots (or suckers if you will) come up from the base that the tree looks more like a shrub. These shoots supply the tree with enough energy to repair the damage. But you probably prefer an oak tree to an oak bush, so I would prune off all but the tallest, best looking sucker. If your tree appears relatively healthy and is putting out some basal suckers anyway, it could be a sign of a more serious problem within the tree.

CHAPTER FIVE

Invasive
Species

What is the definition of *invasive*?

An invasive species is a non-native organism that displaces native species. I often hear people misusing the term. The other day, for example, I heard a woman say that she didn't like Virginia creeper because it was invasive. She may not like Virginia creeper—a valuable native plant in terms of its contributions to food webs—but a native plant cannot, by definition, be classified as invasive. I'm sure she was referring to the fact that Virginia creeper can be aggressive, growing rapidly when conditions are right. And, indeed, like nearly any vine, regardless of origin, it can grow a great deal in a single season. But here's the key: We never see an entire habitat dominated by Virginia creeper. It always grows in concert with many other natives. Why doesn't it take over the way true invasives do? Two reasons. First, it must compete with other aggressive native vines including poison ivy, various grape species, crossvine, and native clematis. Second, a community of natural enemies helps to keep Virginia creeper and other aggressive native plants from overrunning an area. In my yard alone, for example, 33 species of caterpillars eat Virginia creeper, and deer regularly keep it in check too. A true invasive usually dominates large areas, or, if it's held in check, it is constrained by other invasive plants. A patch of Japanese knotweed near me becomes overlaid by porcelain berry every summer; by August, both are overtopped by mile-a-minute weed. In this space, three invasive species are battling for the same area from which native species have been excluded for years.

Plants and animals have always moved around the planet, so the arrival of new species at our shores is a natural process. So what's the big deal? If the new species are more fit than the species already here, then they deserve to replace them.

I hear the "they deserve to succeed" argument quite often. It is true that species have always moved around the planet, but their rate of movement was extraordinarily slow compared to the rate at which *we* are moving plants and animals around the globe today. This matters: if new species arrive at our shores only occasionally, say once every 1000 years, they would never be numerous enough to swamp the complex resident communities, and their addition to existing communities would be so slow, local native species could adapt to their presence.

Four conditions make the ecological contest between resident species and novel (invasive, non-native) species inherently unfair, if we can anthropomorphize just a bit. First, new species typically arrive in novel habitats today through the actions of humans. And they usually are not imported once, but repeatedly. Consider, for example, ornamental plants from Asia. They are brought to this country and sold by the millions over wide geographic areas for decades. The scale of such an influx of new species into our ecosystems in no way resembles any type of dispersal that happened through natural mechanisms in the past. Second, we are importing thousands of new species all at the same moment in ecological time; in any given place today, our native flora must simultaneously compete against dozens of introduced species. Third, in nearly every case, we humans have disturbed native plant communities with our backhoes and bulldozers before introduced species were able to establish successfully. Most introduced species would never have been able to out-compete native plant communities if those communities had not first been gutted by human "development." Finally, new species are typically introduced without the predators, parasites, and diseases that keep them in check in their homeland, so they are healthier when they enter into competition against native plants that must survive attack from multiple species of herbivores and diseases. To call introduced plants "more fit" and thus "more deserving" of a place on this continent than our native plants seems a stretch when we have so unevenly stacked the playing field against our natives.

Why don't we just let nature take its course to control invasives and deer?

The natural world we encounter today is not the same world we would have encountered 500 years ago, 100 years ago, or even 20 years ago. When I was a boy (60 years ago), I could count on finding a box turtle, a spotted turtle, and several species of salamanders any summer day on a walk through a woodlot near my house in New Jersey. I would have kicked up scores of grasshoppers in the meadow I crossed before I reached that woodlot, and I never worried about picking up a deer tick, because I didn't know what a deer tick was; I had explored those fields and woods for years without ever encountering one. If I saw a white-tailed deer, it would have been a very special day, for they were rare where I grew up. I didn't know much about plants, but nearly all of the species I walked by in those days were native to North Jersey. I still remember my first encounter with invasive multiflora rose shortly before I moved from the area. I ran full speed into this new bush I had never seen before and got sliced to ribbons for my troubles. Little did I know that the invasion of multiflora rose was just the first of many more to come, as the ornamental plants used in the surrounding yards escaped cultivation and took over my early stomping grounds.

Were I to ramble behind my old neighborhood today, I would find no box turtles, and the woodland pond that was home to spotted turtles and the breeding ground for local salamanders was long ago filled in with old refrigerators and air conditioners. I would not be able to spend an afternoon in the woods without meeting several deer and the deer ticks they support, and most of the understory shrubs, trees, and vines would be non-native invasives, including those thorny multiflora rose bushes.

As I speak to the public about why these ecological changes are a primary cause of the declines in native species that regularly make headlines today, and how such declines threaten ecosystem function and thus the production of our life support, I am often asked why we don't just let nature take its course to bring white-tailed deer populations back below destructively huge sizes; to control the tick populations and the Lyme disease they transmit that have swelled along with the deer; to control the invasive plants that have escaped our gardens to run amok in our natural areas, replacing native plant communities as they spread. In other words, why can't we let nature fix all the ecological trauma we humans have unleashed on Earth's biosphere? This question goes hand in hand with a related query: Is it wise for us to interfere with natural competition among plants and animals? After all, if invasive plants are more competitive than natives, better at living in North America than the plants and animals that evolved here, shouldn't we just sit back and let them win?

Quite simply, we can't let nature take its course, because it is no longer able to. We have removed the means by which nature used to restore balance in disturbed ecosystems. In the case of invasive species, we have created *unnatural* competitive interactions by importing plants, insects, mollusks, reptiles, birds, mammals, and diseases from other continents without any of their natural enemies. This gives invasive species a huge competitive advantage over our native species whose populations *are* tightly regulated by dozens or hundreds of natural enemies. Throughout most of the country, we have eliminated the wolves, cougars, and bears that used to keep deer populations in check. In fact, unnaturally high deer populations further tip the balance against native vegetation, because deer prefer to eat native plants over Asian species. This has become such a serious issue that US Forest Service botanist Tom Rawinski claims, with much evidence, that unregulated deer populations are a greater threat to eastern forests than climate change.

As to whether we have the ethical right to decide whether a plant or animal lives or dies, I could easily argue that we do *not* have that right. Although it would be hard to find someone who did not believe humans are superior beings and therefore more important than any other species, I can assure you that lions, toads, houseflies, and earthworms feel the same way. The reality is that even those who question our inherent right to manage invasive species make decisions about what lives and dies every day. Every time you choose to plant a non-native species over a native plant, you are deciding whether the chickadee parent in your yard can feed its babies or watch them starve to death because there aren't enough caterpillars to feed them. Every person who installs or continues to maintain acres of lawn is inadvertently deciding whether hundreds of species of native plants and the thousands of animal species those plants support will live in that space or whether they will be replaced by one species of European grass. Every person who hires Mosquito Joe to fog their property is condemning to death not just mosquitoes, but all of the butterflies and bees on their property as well.

The real question is this: on what basis should we decide how to manage our landscapes? I think the best approach is to base such decisions on how they will affect the greater good. For example, if I plant porcelain berry, a highly invasive Asian ornamental, I have inadvertently sentenced generations of native plants to death, because its seeds will escape from my yard and the resulting plant will overgrow just about anything in its path. The same applies to other invasives such as privet, buckthorn, cheatgrass, Himalayan blackberry, burning bush, autumn olive, kudzu, Amur honeysuckle, Japanese barberry, Callery pear, and more. Invasive plants are ecological tumors that spread continuously unless checked by us or some environmental factor. And everywhere they spread, they are killing the native species that run our ecosystems. If you are diagnosed with a cancerous tumor, I doubt if you would ponder whether it is ethical to kill those tumor cells in order to save your healthy cells.

Invasive Plants

Why do some introduced plants become invasive while others do not?

Invasive plants share common traits that encourage their rapid spread at the expense of native plant communities. The most common means of spread is through birds. Invasives such as autumn olive, bush and Japanese honeysuckle, porcelain berry, and so many others all make berries, which birds eat before pooping out the seeds some distance away from the source, often in a natural area. The seeds germinate and the competition with native plants begins. But the deck is stacked against native species, because white-tailed deer, which are overabundant nearly everywhere, don't eat these Asian plants, but they do eat our natives. So the invasive is left to thrive, while the young oak, hickory, tulip tree, witch-hazel, viburnum, birch, and others are eaten as soon as a deer finds them. Natives are also at a competitive disadvantage because they host thousands of species of insect herbivores and diseases, while the invasives host few to none.

The interaction between birds and many invasives confuses people, including land managers. When we see a bird eat a berry, we logically jump to the conclusion that the bird likes the berry and benefits from its nutrition. If that's the case, many people reason, what's so bad about these invasives? They are feeding the birds! But there are two ecological holes in that reasoning. First, no bird species in North America can live or reproduce on berries. Most depend on high-protein seeds or insects to get them through the year, and the vast majority require insects and spiders to reproduce. Berries are just diet supplements. And that's the rub: invasive plants support very few insects, so when an invasive moves in, insect herbivores move out. Moreover, the fact that a bird ate a berry does not mean that the berry gave the bird the nutrition it needed. Most berries mature during fall migration, an event that demands

Warblers prefer berries high in fat, such as these poison ivy berries, to fuel their fall migration.

that birds consume high-fat foods to create the tremendous amounts of energy they need to fly long distances. Fats contain twice as many calories as sugars. Fall is also the time birds that do not migrate need to store lots of fat to make it through the winter months. Susan Smith Pagano at Rochester Institute of Technology has done a series of experiments that clearly show two things: the berries produced by Asian invasives are extremely low in fat (only 1–3 percent) but are high in sugar, whereas native berries such as those produced by dogwood, viburnum, poison ivy, red cedar, and Virginia creeper are high in fat (nearly 50 percent). A classic trait of an Asian plant invasion is that it eliminates native berry makers. So birds eat the high-sugar berries, not because they are good for the birds, but because they are the only berries available. In fact, Pagano has shown that when both native and non-native berries do grow in the same place, birds overwhelmingly seek out the native berries. Birds know what's good for them.

Some invasives move around by means other than birds. For example, Japanese stiltgrass seeds are very small and can travel great distances on the wind, Japanese knotweed is moved by water along streams and rivers, and Norway maple seeds are wind dispersed.

How do you recommend I get rid of invasive plants?

No matter what species you're dealing with, you have to kill the root stock to be rid of the invasive once and for all. Methods that don't work include annual mowing or any form of cutting off the plant a single time. All of these plants rapidly regenerate from the roots, and in a year you won't notice that you ever cut them.

You can kill the roots in one of four ways. Cut them back so often that you exhaust the energy stored in the roots (and by often, I mean at least once every three weeks during the growing season, sometimes for years, depending on how large the root stocks are). Or you can cover the plant with heavy black plastic to block out any sunlight. This can work with invasive grasses, but it often takes many months to kill the roots. A third approach is to whack out the roots with a mattock. What a great invasive remover! Two or three whacks just beneath the base of a multiflora rose bush, a young autumn olive, or the octopus-like center root of a porcelain berry vine, and out pops the root mass. This is an ideal tool for those who want to avoid herbicides.

A mattock is a wonderful tool for whacking out woody invasive plants.

The final way to kill invasive plants is with herbicides. Herbicides can be misused and overused, but when properly used, they are an important tool in our ecological toolbox and can facilitate the fight against invasive plants. I consider them plant chemotherapy. The chemicals we use to fight cancer are nasty but necessary, and they have to be used properly. Not controlling invasives causes far more ecological damage than using herbicides correctly. For annual invasives such as Japanese stiltgrass, a pre-emergent herbicide applied in early spring will prevent the seeds from germinating. Be careful, though, because pre-emergent herbicides will prevent *all* annuals from germinating, not just your target plant, so don't spread it where you are trying to encourage good natives like *Bidens aristosa* (ditch daisy) or *Oenothera biennis* (evening primrose). You can spray the foliage with a product containing glyphosate that translocates the toxin down to the roots through green leaves, but it is often difficult to avoid hitting nontargeted species when you spray. Moreover, recent studies suggest that the surfactants in foliar spray can be deadly to bees. The approach I use, particularly for woody invasives that are too big for a mattock, is to cut the stem or trunk near the ground and paint the stump with an oil-based herbicide that kills the roots. It uses very little material and often works with a single application.

Why is Amur honeysuckle considered invasive? The plants in my yard haven't moved in 20 years!

Ah, but their children have, and their grandchildren have, and their great-grandchildren have moved far and wide from your property! Each of your honeysuckle plants makes thousands of seeds every year, many of which are eaten by birds and then pooped out unharmed a good distance from your yard. Because insects and deer won't eat Amur honeysuckle (*Lonicera maackii*), most of those seeds germinate and grow into new plants. Vast areas of mid-Atlantic forests are now blanketed by this plant, which is a problem because Amur honeysuckle and other invasive ornamentals like it do not produce the caterpillars that are the bread and butter of terrestrial food webs.

What states prohibit sales of invasive plants? How does this work, and why wouldn't every state do this?

Many states have banned the sale of at least some invasives. California, Delaware, Florida, Illinois, Indiana, Maine, Maryland, Massachusetts, Missouri, North Carolina, Oregon, Pennsylvania, South Carolina, and I'm sure others have done so. In some cases, Delaware being a good example, the bans are fairly comprehensive and cover most of the invasives commonly sold in the nursery trade. In most states, though, just a few species are banned, leaving many more to be legally sold in nurseries.

The reason all states haven't banned all invasives is simple: the horticultural trade is very effective at lobbying. The nursery industry argues that the public clearly likes these plants, and if they are banned, nursery owners will lose lots of money. But I question the reasoning here. It assumes that if an invasive plant is banned, the public won't buy an alternative that is not invasive. There is no data to support that contention. There *is* a problem when the decision is left to the nursery owner to stop selling an invasive species voluntarily. If one nursery stops selling a particular invasive but a nearby nursery continues to sell it, the ethical nursery owner will lose business as long as the public remains ignorant about the ecological harm these plants cause. I was on the commission that successfully banned invasive plants in Delaware (with 100-percent support from both sides of the aisle, I might add). The commission heard testimony from several nursery owners, and I was surprised to learn that many *favored* a ban, because if no nursery is allowed to sell a plant, no single nursery loses any business. Customers will simply buy another plant. In fact, as homeowners learn more about how biodiversity declines impair the production of the ecosystem services we all depend on, they are buying *more* ecosystem-friendly plants, not fewer, replacing lifeless lawns with plants that enhance ecosystem function rather than degrade it. There are 135 million acres of residential landscapes in the United States. If we all add more non-invasive plants to our properties, it will not hurt the nursery trade!

If you remove the invasives, what's to keep them from coming back?

You are! Our fight against invasive species is never ending, which is why it is such a devilish problem. A constant "seed rain" of invasives comes from your neighbors, and the soil holds a vast seed bank. The good news is that once you have eliminated the monsters—the autumn olive and multiflora rose and Amur honeysuckle with 6-inch-diameter trunks—the new plants that come in each year will be tiny seedlings that are easily whacked out with a small hand mattock. Moreover, as your native replacements begin to grow, they will slow the invasion considerably by throwing shade. Remember that you are responsible for managing invasives on your property only. And 78 percent of the entire country is privately owned (85.6 percent east of the Mississippi is privately owned), so if we all removed invasives from our own properties, we would be well on our way to achieving effective control. The seed rain would diminish over time, leaving us more time for happy hour each spring!

How do I remove or destroy buckthorn, honeysuckle, and other woody invasives?

Your goal with all invasives is to kill their root systems. Just cutting the plants at the base won't do it. The two best choices are to whack out the bigger buck-thorns, autumn olive, and bush honeysuckles with a full-sized mattock. It is a lot of work, but it is effective and fast. Two or three whacks will pop an 8-foot buckthorn right out of the ground. Your other option is to saw the plant at its base and paint the stump with an oil-based herbicide. Note that I do *not* recommend glyphosate for woody plants; it is designed as a foliar spray, not a stump treatment.

Japanese honeysuckle and oriental bittersweet vines are even easier to kill. When vines climb, they usually do so via one or two main stems. Locate these, snip them off close to the soil, and paint them with an oil-based herbicide. This will kill the roots and you're done. You can have fun pulling the vine out of a tree, but that's not necessary. Invasive vines decompose quickly after they die. If you don't want to use an herbicide, snip off any new growth as soon as you see it. You'll have to do this for a few years until you exhaust the energy stored in the roots.

How does one dispose of invasive plants after they are removed?

There are several options for getting rid of invasive plant material, but some are better than others. The goal is to kill any plant parts that remain alive after you cut or pull the plant. Air drying works well for most species and is particularly effective for multiflora rose, porcelain berry, bush honeysuckle species, privet, oriental bittersweet, and kudzu. You can cut them and create brush piles that decompose remarkably fast. Other species, particularly some woody invasives such as autumn olive, Russian olive, privet, bamboo, and buckthorn, are harder to snuff out, and you have to make sure none take root while you are waiting for them to die. Early on in the restoration of our property, I whacked out several autumn olive trees with my mattock in June and then cut them into stakes that I planned to use as anchors for deer cages in fall. The stakes were stored in the sun all summer before I pounded them into the ground to support my cages. To my astonishment, nearly every one took root!

Disposing of plants that have gone to seed is more challenging, and that's a good reason to remove invasives before they produce propagules whenever possible. Many seeds can continue to mature even after the parent plant has been cut. This is true of oriental bittersweet, mile-a-minute weed, privet, multiflora rose, miscanthus, and other invasive grasses like Japanese stiltgrass— most plants that set copious amounts of seeds or berries. Some recommend bagging these plants in large plastic garbage bags, but if you're dealing with even a moderately sized invasion, that can mean a lot of bags and a lot of plastic. Burning works if you have the space. Putting the plants in a garbage can to be collected is not a good idea, because the seeds are sure to spread far and wide. And then there are plants like Japanese knotweed that can produce a new plant from a section of root as tiny as your fingernail. Piling these plants up and burning them or covering the pile with heavy black plastic to bake in the sun is an effective approach.

We bought a 100-year-old house on a small urban lot that has been landscaped with several very well-established invasive ornamentals (Japanese maple, burning bush, Norway maple). We realize they serve little ecological value, but we are not sure if it is worthwhile to cut them down and replace them with oaks or other natives. Thoughts?

Removing burning bush is definitely worth the effort, since it is one of our very worst invasives. Its seeds are not staying on your property and are moving into nearby natural areas, where they germinate and then spread unabated. Your Norway maple is taking up a lot of space, but unless it is within seed-blowing range of a natural area, it probably poses little risk of spreading. Japanese maple is the least of your worries. Although I occasionally find one on my property that must have moved here from at least half a mile away, it is not high on my list of problematic invasives. But keep in mind that each one of the non-natives is more like a plastic statue than an ecologically functioning tree—so the question really is, how many statues do you want in your yard?

How can I control Japanese stiltgrass?

Japanese stiltgrass has become an omnipresent annual over much of the mid-Atlantic region, and despite its dying back each winter, it is one of the toughest invasives to control. Mowing won't do it because stiltgrass, like other grasses, produces seeds in two places: at the terminal ends of the blades and, incredibly, in the blade axils down very low! You cannot mow low enough to destroy them. So stiltgrass is very tough to control without herbicides, and you have two options. Pre-emergent herbicides, when spread as granules in the early spring, will kill germinating seeds without hurting existing plants. Of course, pre-emergents will kill *all* species of annuals, not just stiltgrass—a big downside. Another option would be to use a grass-specific herbicide, which will kill only grasses, natives included. What's really needed for this stubborn invasive is a species-specific disease that will kill all stiltgrass as effectively as the chestnut blight killed all of our American chestnuts.

In writing about invasives, I always come across the phrase, "the seeds are distributed by birds and mammals," during my research. As I understand it, local bird populations frequent a fairly small range of habitat. So if a homeowner has a barberry shrub in her yard, and she doesn't see it taking over, should we assume that the seeds of the shrub have been distributed by birds to a forest miles away? This is a particular bone of contention with homeowners, and I'm wondering whether I fully understand the dynamic.

Two things are going on here. You're right: local resident birds do have a fairly small home range. But what about those fall migrants that eat a berry and then fly up to 300 miles in a single night, pooping seeds as they go? And don't forget the mammals. If a dispersing buck eats a barberry fruit and then decides to do his dispersal run, he can go 10 miles before he adopts a new territory. All it takes is one of those events to start a new local population that will spread slowly each year like an ecological tumor.

This summer I took on the gigantic task of eradicating Japanese honeysuckle from the woods behind my house and my neighbor's house. In doing so, I have run into several snags. First, in removing the honeysuckle, I have left the woods open to other invasives taking over. I can't plant native species in the huge amount of space that is now empty. I was wondering if there is something else I should be doing after I remove the honeysuckle? Second, I am having difficulty identifying species of plants. I don't want to pull everything in my backyard and then start over from scratch. I was wondering if you have any suggestions (books, websites, and so on) that could help guide me in separating the plants that should be pulled versus the ones that should stay?

First, do you have a deer problem? Even just a few deer will eat many of the natives that sprout in your yard, making it look like nothing but invasives grow there. For now, let's assume you do not have a deer problem. One way to repopulate your yard without planting anything is "addition through subtraction." You just keep pulling the bad guys while leaving the good guys. But, as you say, you need to recognize the good guys.

No matter where you live, the interest in native plant gardening is exploding, and how-to books are available for the Southeast (*Native Plants of the Southeast: A Comprehensive Guide to the Best 460 Species for the Garden*), Northeast (*Native Plants of the Northeast*), Southwest, Texas, and Rockies (*Landscaping with Native Plants of the Southwest, Landscaping with Native Plants of Texas, Native Plant Gardening for Birds, Bees & Butterflies: Rocky Mountains*), Midwest (*The Midwestern Native Garden*), California (Calscape), and the Pacific Northwest (*Gardening with Native Plants of the Pacific Northwest*). Another easy method is to take a good picture of a leaf or the whole plant with your phone and post it on iNaturalist, asking for identification help. You'll usually get an answer quickly. Speaking of iNaturalist, the organization offers another free app, Seek, which is quite good at plant identification and is well worth trying. If you do have a deer problem, you might put wire cages around the plants you want to keep until they grow large enough that browsing deer can't kill them.

I just got home from a week in West Cape May, New Jersey. The front yard of the place where I was staying had a Siberian elm. This tree, at all times during the week, was full of warblers. We shook a branch, and all these teeny flies (like gnats) flew up. I'm happy the tree species is not aggressive and coming up everywhere else, but apparently the birds have found a goldmine in this tree in Cape May. What do you think? Are the insects adapting to non-natives and the birds are beginning to benefit?

What you have experienced is a little bit of Eurasia in a New Jersey yard. When Siberian elm was brought over for the ornamental trade, along with it came the cockscomb gall aphid, a specialist on that tree. The little gnats you saw are aphids that fly to the tree, where each lays a single egg, which then spends winter on the tree. The birds have figured out that even though aphids are tiny, because there are so many of them, it makes a visit to the elm worth-while. So, no, our insects are not adapting to the elm: the aphids arrived here already adapted! Warblers in California visit the tipa tree (*Tipuana tipu*), an invasive plant from South America, for the same reason. When it was introduced here, a tiny cicada-like psyllid that is abundant on this tree came along for the ride. Introduced aphids on Norway spruce supply North American birds with snacks as well. Birds do not care whether an insect is native or not; they will eat it when it's convenient.

Should I remove the white non-native mulberry in my yard and plant red native mulberry instead?

That is exactly what I did when my wife and I moved into our new home in 2000. A large white mulberry grew on the edge of our property. It wasn't native so I took it down, and I have been sorry I did ever since. Here's why: Through a genetic process called introgression, when white mulberry crosses

An invasive aphid from Europe forms cockscomb galls on Siberian elm.

with red mulberry, the resulting offspring are all white mulberry. White mulberries have completely replaced red mulberries throughout the entire eastern half of the country. A botanist in a position to know told me recently that you have to go to Texas to find a real red mulberry. Ecologically speaking, red and white mulberries are practically identical. They both make copious amounts of fruit that's high in sugar during early summer, exactly when birds need carbohydrates after long months of feeding their nestlings on insects high in protein and fat. What typically differentiates a native from a related non-native plant is its ability to support lots of insects. Not so with mulberries, because *neither* mulberry supports many insects. By taking down the white mulberry on our property, I eliminated the only source of summer berries for our local birds. Now I leave white mulberries that have seeded into our property, but it has been 22 years since I cut down the big one, and all of the young ones combined still don't come close to producing the berries that single large tree produced. So my recommendation would be to leave your white mulberry where it grows. Replacing it with a red mulberry won't enhance your local food web at all.

I have heard and read a fair amount about the invasiveness of Pugster dwarf butterfly bush's big brother, buddleia. The biggest condemnation called it "cocaine for butterflies" because they love it so much they get drugged into laying eggs there, and then the caterpillars can't survive on a non-host plant. I wonder if retaining these pretty plants is irresponsible.

I think someone has stretched the truth a bit. I have never heard of a butterfly laying eggs on butterfly bush because it was too "drugged" to leave. The point is that butterfly bush does not support larval development of any butterfly east of California. It gives adults nectar, hence its name, but it does not make more butterflies by nourishing caterpillars. My problem with butterfly bush is its invasive tendencies. If the Pugster makes viable seed, then it could also become a problem in the future. Butterfly bush is a major invasive in the Pacific Northwest, in parts of eastern Pennsylvania, in Hawaii, and in other areas. If you can grow Pugster without it spreading around, I have no problem with it.

I get the idea why native plants are better to plant than non-natives. But I do see a lot of activity on the non-natives I have in my garden. For instance, the huge burning bush I have is loaded with birds, it appears mostly robins, all going for the berries. Also butterfly bush. All season it's been loaded with butterflies. Is this food source bad for them? They certainly seem to love it.

When evaluating the ecological value of a plant, one needs to look at the net benefit (or loss) from having this species in the garden. For example, burning bush makes lots of berries that birds eat, and then the birds poop out their seeds in our natural areas. This is what makes it so invasive. Its berries are a plus, but birds do not live on berries alone. They rear their young on insects, and burning bush does not support the thousands of caterpillars birds need to reproduce—a big minus. Burning bush also spreads into natural areas, displacing the native plants that *do* produce the insects that birds need. That's another huge minus. So, in this case, the ecological minuses of having burning bush far outweigh the pluses. A landscape full of burning bushes has no breeding birds, berries or not. Burning bush, therefore, is a disaster for our bird populations.

What are some good replacements for butterfly bush in a city garden?

You have many native alternatives. Joe Pye weed, both *Eutrochium purpureum* and *E. dubium*, are some good tall and shorter options, respectively. Sweet pepperbush, buttonbush, and Virginia sweetspire are all midsummer bloomers. Plant catalogs will tell you that these plants like wet areas, but they'll also do well in drier soils. I would combine as many of these as you have room for to get longer bloom time. You might also consider bottlebrush buckeye, which is a wonderful butterfly magnet.

How do we prioritize which invasives to get rid of?

There are different ways to prioritize which invasives to attack first. Some people start by choosing the easiest one to deal with. Pulling out garlic mustard, for example, is easy, although it's ineffective because it disturbs the soil and therefore activates the seed bank. The end result is more garlic mustard the following year. I like to attack the species I know to be most harmful (that spreads the fastest, occupies the most territory, changes the soil the most, throws the most shade, feeds the fewest insects, and so on) and those I am most likely to remove successfully. Japanese stiltgrass clearly falls into the most harmful category, but because of the seed bank issue, it is a species I am unlikely to remove successfully. There is nothing worse than investing weeks of time and energy into removing an invasive, only to see no difference the following year. So I typically direct my energy toward woody invasives like autumn olive, burning bush, multiflora rose, porcelain berry, Amur honeysuckle, and others. When I kill a number of these, I can really see results. I pile up the bodies into a huge mound and get that feeling of accomplishment that I was born craving.

What is the best way to get rid of bishop's weed?

Bishop's weed, also known as goutweed, is an invasive ornamental from Eurasia that has been spreading from our gardens as a thick groundcover since 1863. It is banned for sale in Connecticut, Delaware, Massachusetts, and Vermont but is still sold elsewhere. That would stop on a dime if the nurseries that sold it were responsible for its control once it escaped. As you might have guessed, bishop's weed joins a group of other invasives including lesser celandine, phragmites (common reed), and Japanese knotweed, which are all maddeningly difficult to eradicate. Bishop's weed is stoloniferous and sends roots laterally in all directions. You can pull it up, but if you miss any one of the lateral roots (and you will), it will be back in short order. Like all plants, you can kill it if you cut every stem every week for one or two years, never letting it photosynthesize to revitalize its root system, but that labor-intensive approach is a nonstarter for most people. Covering it in black plastic can starve it of sunlight, but only if you cover every bit of it for at least two years. Of course, you can spray it with an herbicide, but collateral damage on nearby natives is difficult to avoid. Choose your poison, as it were. Controlling bishop's weed ain't easy.

Can pines be invasive?

I don't know of any true pine that is invasive in the United States, despite the fact that we have imported many species of exotic pines for ornamental purposes. We do have a serious invasive problem with several *Casuarina* species, commonly called Australian pines, in Florida and the West Indies, though these are angiosperms, not gymnosperms. In other parts of the world, however, non-native pines have become destructive invasives. Monterey pine, for example, has spread throughout the southern hemisphere. It has become particularly problematic in South Africa, where it displaces native species and sucks up so much water from the water table that streams have actually disappeared from the landscape.

I have been religiously culling the invasion of Amur honeysuckle we have. But in the process I have unearthed a massive seed bed of garlic mustard; it is now everywhere the honeysuckle was! I am currently using a mixture of glyphosate and hand pulling; any other suggestions to help me get ahead of it?

Check out Cornell University Professor Bernd Blossey's work on garlic mustard. After 30 years of plot treatments comparing various ways to eliminate garlic mustard, believe it or not, the treatment that produced the best control was to do nothing! Apparently, garlic mustard reproduction begins to fail after several years if the soil is not disturbed. Every time you pull one or disturb the soil in any way, you reset the clock.

What do you think about using classical biological control agents for invasive plant species? Cypress spurge is one of the main invaders of a property I'm managing as a significant pollinator habitat; without the extensive use of herbicide, control is quite unlikely. I've suggested the use of biological control agents, but members of my organization's board are understandably wary. The limited research I was able to find indicates that their effects, both direct and indirect, can be highly variable based on the control agent and the strength of the relationship. So I'd like to know your thoughts about the pros and cons of them.

Biocontrol, the importation of specialist insects to control an imported pest, is one of the best tools in our ecological toolbox for control of invasives. Your board has no justified basis for being wary of it in today's world. The only cases where biocontrol went wrong happened decades ago, before there were rules regulating it. People would import generalist predators, parasitoids, or herbivores, which, of course, attacked non-target species. That doesn't happen today. In 1976, strict rules were established regulating what had to happen before any species could be introduced to control pest animals or plants. USDA quarantine facilities were built so that biocontrol candidates could be evaluated for host specificity before being released into the wild. Regulations are strict; it must be experimentally demonstrated that a potential control agent will not attack any native species that are even remotely related to the target pest. Such rigorous testing often takes a decade to complete and to date has prevented imported agents from switching to any non-target species. When the public hears about biocontrol disasters, it is always from introductions that were made before these tough regulations were put into place. It's true that a single imported herbivore often doesn't control the pest successfully; it often takes a whole community of insect herbivores to control an invasive plant. But there are several examples of spectacular successes with single introductions, so it's always worth a try. Several related beetles are being used for control of cypress spurge, for example, and the results are promising. I'm all for it!

Are there two species of bittersweet?

There are two species today, but there may not be in a few years. Oriental bittersweet, *Celastrus orbiculatus*, was introduced through horticulture because of its pretty seed arils. Unfortunately, it has proven to be highly invasive, has spread nearly every place it was sold, and it readily crosses with American bittersweet, *C. scandens*. Through introgression, a genetic phenomenon that displaces the genes of one mating partner, the offspring of such crosses are 100 percent oriental bittersweet. As of this writing, oriental bittersweet has eliminated American bittersweet everywhere except northern Maine.

Aren't invasive plants like privet and autumn olive good for pollinators?

No, they are not good for pollinators for two reasons. First, autumn olive, privet, bush honeysuckle, and many other invasives do produce flowers, and some generalist pollinators such as honey bees and some bumble bees will use those flowers. The problem, however, is that these invasive plants push out all of the blooming native plant species that would have presented pollinators with a sequence of blooms from April to November. So whereas privet blooms prolifically for one week of the year, during the other 51 weeks, no native plants are left to bloom in areas where privet has invaded (that's millions of acres in eastern North America). The sad fact is that when invasive plants move into an area, the amount of pollen and nectar available for pollinators decreases. In addition, non-native plants are not used by specialist bees. More than a third of the 3600 species of native bees in North America can reproduce only on the pollen of plants with which they have coevolved.

I have a tremendous amount of lesser celandine in my garden. We have lived here for 16 years, and the lesser celandine was not present when we moved in. Whenever I am weeding or planting, I try to dig up the plants and throw away the bulblets, but there are a lot more than I can possibly eradicate by hand. I am living with it since it "disappears" as summer unfolds. Should I be concerned?

Lesser celandine is a bugger. It does disappear later in the season, but it competes with other spring ephemerals early in the year. For large infestations, hand digging really isn't a viable option, although a small hand mattock is effective until your wrist gets too tired. You can hit it with glyphosate shortly after the leaves appear in spring (but before flowering) if you so desire, but there are downsides to doing that. Like so many of our impossible invasives, lesser celandine was introduced as an ornamental and is still sold in some places. This is one of our invasives for which there is no silver bullet!

I recently had a large swath of bamboo removed from my property. I am planning to plant some oak trees in the area. My concern is the underlying root system of the bamboo. Will the bamboo roots inhibit the growth of the trees? Is there a way, short of brute strength, to destroy the root system?

The answer depends on how the bamboo was removed. If it was simply cut, the roots were not killed and all of the bamboo will return. If it was killed by herbicide, then the roots *are* dead and their bodies won't bother your oaks. Just make sure you have actually removed the bamboo before planting into the mass of roots. Living bamboo roots will overwhelm anything you try to add.

My husband and I bought a little over seven acres last year. Toward the back of our property is a large patch of reed phragmites. We had two master gardeners come out for consultations to guide us on planting natives throughout our property, and both recommended spraying glyphosate on the phragmites. I eagerly followed their recommendations. Afterward, I began reading and became aware of how hazardous glyphosate is. I also watched the documentary *Kiss the Ground*, which prompted me to research alternatives to prepping a site. Right now, I feel ill-informed about what the "right" thing to do is. Professionals say it's a contentious subject, and everyone weighs the risk of bodily harm (cancer) versus the greater ecological good. A small spot in my backyard that needed phragmites removed seemed like the ideal situation for using glyphosate. As much as I'd like to say that this response gives me peace, it doesn't.

To herbicide or not is always a tough call, and I would need to see your property to lend an educated opinion. I will say that much of what you read about glyphosate is based on continuous overuse in agriculture. When making your decision, you need to consider the costs and benefits together to try and determine whether the costs outweigh the benefits. I liken herbicides to chemotherapy: invasives are ecological tumors, and ignoring them can kill an ecosystem. Spraying is not cost free, but not spraying can also have serious ecological costs that often are barely considered.

The real problems come from misuse of this tool. When used properly, glyphosate does not pose as much risk to you as your toaster does (370 people are killed each year by toasters!). You cannot kill invasive plants without killing the roots. Herbicides are the easiest way to do that, but they're not the only way. If I owned your place, I would want to know how long your patch of phragmites has been there, whether it is spreading or not, how likely it is to spread to un-infested sites, and whether mowing it with a bush hog (or something similar) is an option. Repeated (and I mean repeated) mowing will eventually exhaust the root systems and give you good control with no herbicide. But if the soil is too wet, that may not be an option. I know of some groups who are religiously hand-clipping small patches of phragmites to get the same effect, but that, of course, is a lot of work. Phragmites is difficult to control, even with herbicides, so a single spray is not likely to work. The best-case scenario would be that you have a small, stable patch that is not spreading and is unlikely to move to other sites. In that case, I would put it on the back burner and focus on the other seven acres.

When is the best time to control invasive plants?

If you are using mechanical control (ripping them up by the roots or repeatedly cutting them back) the best time to control invasives was yesterday. If you are using herbicides as a stump paint treatment (my favorite approach), early fall is the best time, because that is when nutrients are being translocated back to the roots. The herbicide will be moved down to the target roots as well. This is why painting with herbicides in spring is far less effective. The upward vascular flow once foliage begins to develop can inhibit or prevent herbicide translocation to roots. That said, oil-based herbicides can also be used throughout winter. If you are using a foliar spray, the best time to spray is during spring/early summer leaf expansion, when photosynthesis is at its peak.

Invasive Animals

How do I get rid of House Sparrows?

House Sparrows are not actually sparrows but a species of finch that was introduced to the United States in the 1850s by immigrants hoping to control pests in orchards. It didn't work! Instead, House Sparrows have become one of our most serious invasive species, threatening native tree-hole nesters like bluebirds across the country. They can be controlled, however, through passive actions or active control methods. One easy approach is to locate bird houses away from human structures. House Sparrows are so called because they love to nest near human structures, especially in agricultural communities where there is plenty of available grain. Another passive approach is to plug the entrance to a bird house if it has been taken over by House Sparrows. This might encourage the pair to seek nesting sites elsewhere, but it has the drawback of rendering your box useless to bluebirds, chickadees, titmice, and Tree Swallows. Your choice of housing material can help as well. House Sparrows don't like nesting in Gilbertson PVC nest boxes, but bluebirds and chickadees don't mind them at all. You can also take steps to reduce the attractiveness of your nesting area to House Sparrows. For example, if you use bird feeders, stick to sunflower seeds, which House Sparrows do not like, and avoid filler seeds such as milo, millet, and cracked corn, which are their favorites.

If you are feeling a bit less charitable toward your House Sparrows, you are within your legal rights to use active (lethal) control measures. Remember that House Sparrows are invasive species and are not protected by law. You can monitor your nest box and remove House Sparrow nests as soon as they are built or after the eggs have been laid. Male House Sparrows will aggressively defend the box and keep other birds from using it, so you may find it necessary to remove the male itself. This can be done by trapping the bird in the box using a Huber, Van Ert, or Gilbertson sparrow trap. Another option that is particularly good if you have a large population of House Sparrows is to trap them outside of the nest box with a repeating bait trap. These traps will catch several birds at once, but they have to be checked often in case some native birds are trapped and need to be released. There are several humane ways to euthanize House Sparrows once they are trapped.

One final note: When my wife and I moved into our new home, which we built on what was formerly farmland, a sizable population of House Sparrows was there to greet us. We did nothing to manage them, but we did start planting trees and shrubs. It didn't take long to transform the open habitat House Sparrows prefer into scrubland and then patchy forest with 70-foot trees, habitat that House Sparrows shun. We have not seen a House Sparrow on our 10 acres for more than a decade.

Do English sparrows and starlings compete with our native birds?

Yes, they do, because they are both cavity nesters. Eighty-five North American bird species nest only in cavities, mostly tree holes originally hollowed

Without a nationwide campaign to erect bluebird houses, we might have lost these beautiful birds.

out by woodpeckers. There are usually insufficient numbers of tree cavities for the birds, particularly in residential neighborhoods, where we tend to cut down dead trees. Both English sparrows and starlings are aggressive species that can easily evict tree-hole nesters including bluebirds and Tree Swallows. If our native birds can't find an unoccupied tree cavity, they lose their opportunity to breed until the following year. That's a big loss for birds that often live for only a few years. In fact, it was the explosion of English sparrow and starling populations that decimated bluebird populations in the first half of the 1900s. If bluebird enthusiasts hadn't started a nationwide campaign to put up bluebird houses, we might have lost those beautiful birds entirely.

I have not seen a Carolina mantis in decades—only the non-native mantids from Europe and China. I have read that the non-native mantids are probably harmful to our native ecosystems because they kill too many native insects (and hummingbirds as well). Is this true?

The Chinese mantid was introduced to the United States near Philadelphia by the nursery trade in 1896. It rapidly spread throughout the Northeast,

The easiest way to control Chinese mantids is to put their egg case, shown here, in the freezer.

continues to spread westward, and is now the most commonly encountered mantid in the East. North America's largest mantid, often exceeding 5 inches in length, the Chinese mantid is a "sit-and-wait" predator, positioning itself next to flowers so that it can grab and eat anything that comes to gather pollen or nectar. And I mean anything, including butterflies (monarchs too), native bees, honey bees, syrphid flies, beetles, small lizards, and, as you mentioned, even hummingbirds. The European mantid is another invasive species in the eastern United States. Though it is similar to the Chinese mantid in appearance, it is always a bit smaller and less common in most places. There is correlational evidence that together these two species have displaced some of the 21 mantid species native to North America. So if we value our native flower visitors and mantid species, it's not a bad idea to remove these creatures when you find them. The easiest way is to destroy their globular egg cases, which are laid on erect vegetation and easy to find during winter. By the way, the urban legend that I grew up with—that you would be fined 50 dollars if you killed a praying mantid—is just that, an urban legend. I have no idea how that got started, but it proves you don't need social media to spread misinformation!

Are yellow jackets, paper wasps, and hornets non-native?

Yellow jackets are native and so are almost all paper wasps, but at least one introduced species is widespread. Hornets are native except for the giant European hornet and, of course, the giant Asian hornet, sensationally called the murder hornet.

Last night I found what I believe is a "murder hornet." The news makes it sound like they are not in this area. Is this something I should worry about? If there is one, there must be a nest somewhere.

If you live anywhere east of Washington State, you have not found a murder hornet, which more appropriately is called the giant Asian hornet. That species is restricted to western Washington State, and by the time this book is published, it will have very likely been successfully eradicated. What you found was a giant European hornet, an invasive species that has been in the United States for more than a century. You can worry about it if you like worrying (you'd be surprised how many people love to worry!), but let's do some risk assessment. Suppose you are 50 years old. In that case, European hornets have been in your life since the day you were born—yet this is the first time you encountered one. That's not because you've been unusually lucky; they just aren't very common in most places. What you probably are worrying about is whether or not it will sting you. If you picked it up with bare hands and shook it, it probably would, but you're not going to do that. In fact, I have never heard of someone being stung by a European hornet. I'm sure it's happened somewhere, but it must be very rare. So, based on all of that, the risk of you actually being harmed by a European hornet is very, very small. European hornets are, however, invasive species that eat our ecologically valuable caterpillars all summer long. Every caterpillar they eat is lost as food for some hungry native bird.

Do Asian lady beetles have any redeeming value?

I guess Asian lady beetles' value depends on your perspective. *Harmonia axyridis*, also called multicolored Asian lady beetles, were introduced here in 1916 to help control the soybean aphid. They did a good job! But like so many other non-native insects, they have no native natural enemies and soon

Polistes dominulus is an introduced paper wasp that may be displacing native paper wasp species.

The giant European hornet is more common and smaller than the giant Asian hornet.

spread throughout the United States. The upside is that they do eat lots of aphids. The downsides are that they occasionally explode in numbers, and, like other lady beetles, they overwinter as adults in protected areas. And the protected areas Asian lady beetles prefer are our houses. They can gather in huge numbers in the upper corners of a ceiling, behind picture frames, in an attic, really anywhere they feel comfortable. This is nothing more than a nuisance, but ecologically there are downsides as well, for Asian lady beetles have competitively excluded native lady beetle species wherever they have gone. Some native species have declined to the point of alarm for conservation biologists.

Can birds control winter moths?

Winter moths were accidentally introduced into the Boston area in the 1990s, where they caused widespread defoliation of forests and shade trees. Birds love this geometrid inchworm, but in residential areas hard hit by the winter moth, there haven't been enough birds to control them. Enter Dr. Joe Elkinton of the University of Massachusetts. In 2005, he initiated a biocontrol program using the specialist tachinid fly *Cyzenis albicans*, a European parasitoid that attacks only winter moths. For 14 years, he released several thousand tachinids in areas with high winter moth populations and the results have been phenomenal. From Maine to Connecticut, winter moth populations have all but disappeared.

So much information is out there about the perils of the Asian jumping worm and its adverse effects on the soil, native plants, and forests. Most information suggests there aren't any ways to eradicate them that won't also harm other beneficial creatures. What can urban gardeners do to help their native plantings survive in an environment where this worm is present?

There is not a lot of good news on the subject of Asian jumping worms. They have been in the United States for quite a while, with the first records as early as the 19th century. As with so many other serious pests, they were brought in with nursery stock from Asia. But they spread very slowly on their own, and infestations were small and few until about 30 years ago when they exploded, again through movement of nursery stock, into the eastern United States in areas with enough moisture to support them.

The ecological impact of earthworms is a confusing tale. Charles Darwin studied European worms (*Lumbricus terrestris*) in detail and concluded that they are, on the whole, beneficial to soil biota because they aerate the soil, help recycle leaf litter, and thus add nutrients to the soil in a form that is available to plants. When European earthworms were brought to the United States, they behaved themselves and added important services to our eco-systems, particularly in the northern parts of the country where native earthworms had been eliminated by the glaciers.

But all species of earthworms are not created equal. There are at least three species of Asian jumping worms, so called because they are very active, and when they are brought to the surface, they wriggle around so much that they appear to be jumping. They basically do what European worms do (eat leaf litter and other organic matter in the soil), but they do it much faster. This is definitely a case of too much of a good thing. They grow twice as fast as European worms, they are parthenogenetic (reproduce without sex) and so increase their populations way too fast, and they don't just help decompose leaf litter, they eat *all* of the litter and turn *all* of the topsoil into castings that are easily broken down by rain, which then washes away the nutrients within.

The challenge has always been to kill the worms without killing the thousands of other creatures—the *Collembola* springtails, proturans, nematodes, centipedes and millipedes, snails, fungi, and bacteria—that contribute to healthy soil communities. Your best chance to control them is when you first discover them. They overwinter as tiny egg pods and are most vulnerable when they first hatch in spring. An organic fertilizer made of tea seed meal is commonly used on golf courses and apparently is effective in killing young jumping worms. If you can't find a distributer of tea seed meal, ask your local golf course where they get it.

While collecting neighbors' leaves, I noticed under the bags large, active worms with a white band. Through my research I have discovered that they are jumping worms. Our extension office suggests I solarize them.

These worms are now everywhere. I usually have scissors in my garden bucket. Can I just cut them up and mix them back into the soil?

Cutting them (if you can get them sit still) won't do them any good, but it won't guarantee that they will actually be killed. Putting them in a plastic bag and leaving it in the sun *will* kill them, however. The more adults you kill, the fewer egg sacs they will leave in the soil.

In your books, you note that some insects, and in particular some caterpillars, will eat only one species of plant. With regard to an invasive plant species, if there was an insect species native to the invasive plant's home range that was similarly dependent on only that one invasive plant species, would you support importation of that insect to combat the invasion? For example, to attack kudzu or purple loosestrife.

Yes, I would support it. In fact, that is what classical biological control is all about—introducing an insect from the home range of the invasive plant to help keep it under control where it has escaped cultivation. But there are important caveats to my support. The controlling agent has to be truly host-specific on the invasive plant. To determine this, potential biocontrol agents must be tested within quarantine facilities on all native or agricultural plants that are even remotely related to the invasive. Kudzu is a good case in point. We would love to find a good biocontrol agent that could keep that plant from smothering the South. The problem is, kudzu is a close relative of soybeans, and every insect that has been considered for control of kudzu also attacks soybeans. Consequently, permission to import these insects is always denied by the USDA. But purple loosestrife, your other example, shows how well biocontrol *can* work when executed properly. Two species of chrysomelid leaf beetles have been imported from Europe to control purple loosestrife— and control it they have. They are host-specific on purple loosestrife (won't eat anything else), and in most infestations where they have been released, excellent to total control has been achieved.

Do invasive species harbor any plant diseases or pests?

Yes, they can, although there has been little study of this. One notable example is English ivy, which serves as a reservoir for bacterial leaf scorch, a serious disease of oaks. Buckthorn, a serious invasive in our northern states, is the alternate host for soybean aphid, a significant agricultural pest. I'm sure there are other examples as well.

I need to know what the impact of losing most of our spring ephemerals due to an invasion of lesser celandine will have on my area of Mid Michigan. I have been given approval by the local parks department to attempt a massive ecological restoration on the Grand River. This includes experimental ways to eradicate this difficult species and to experiment with transplanting and propagating the threatened ephemerals.

I am very sorry to tell you that I don't know of a silver bullet, or even a bronze bullet, to help with lesser celandine. It's just like Japanese stiltgrass: we don't know how to eradicate it. We need a species-specific disease to attack these plants, and I don't think anyone is working on that. I can tell you that hand pulling won't work. I supposed you could spray the devil out of it, but what about the seed bank? Smothering slows it down but does not kill it. It will come up again the following year. I wish I had better news.

For the past two years, our area has been severely affected by the spongy moth. I'm wondering if it is OK to spray BT next year or if it is harmful to other important moths and caterpillars.

This is a judgement call. Your best bet is to understand the facts. First, BT (*Bacillus thuringiensis*) is a naturally occurring bacterium that will kill *all* the Lepidoptera on your oaks, not just spongy moth. Next, a big spongy moth population one year does not necessarily mean it will be big the following year. When we have cool, wet springs, a fungus that controls spongy moth becomes active and can just about eliminate the population for several years. New England is due for this fungus to outbreak. All it takes is some rain. So I would wait until egg hatch next year and then see what conditions are like. I guess what I'm saying is this: spray according to conditions, not according to the calendar. Finally, serious defoliation two or three years in a row can kill an oak, especially under drought conditions, and we don't want that. Bottom line, spray BT when necessary, but consider all factors to determine whether it's necessary.

I have heard there is a fungus that controls spongy moth. Can we buy and spray this?

Yes! Dr. Ann Hajek of Cornell University discovered a fungus (*Entomophaga maimaiga*) that made its way to North America by unknown means a few decades ago, and it's now the most effective control for spongy moth. Its spores reside in the leaf litter under an oak tree (another reason to retain leaf litter under your trees). In spring, when spongy moth larvae hatch and balloon (that is, release a silken thread and let the wind take them where it may), many land on the ground and have to crawl through leaf litter to get to your trees. This is one time they can become infected by the fungal spores. Spongy moth caterpillars can also become infected in later instars (larval stages of development) because they have a habit of crawling down tree trunks during the day to hide in the leaves beneath a tree.

My village documented the presence of spongy moths last year with sightings and traps. They are deciding whether to adopt an aerial spraying service to apply BT to most of the community. The village staff got much of their information from the Department of Agriculture. It is their understanding that this moth's larval stage/caterpillar emerges before that of most native species. The spraying will be conducted in a small window of time dictated by the oak's leafing stage and the moth's emergence. Is this the best approach to protect the village's oaks?

How big is the spongy moth population at this point? How many adults were trapped? Has anyone found any egg masses? If the population is large, which is doubtful if this is the first year they have been detected, then spraying might be justified. But if the population is tiny, the BT spray will kill all of the early season Lepidoptera, which will just about eliminate your woods as stopover sites for migrating birds or breeding sites for residents. Spongy moths do emerge early, but many other species, particularly the geometrid inchworms that fuel bird migration, are early season species as well. In fact, many geometrids overwinter as caterpillars, so they would certainly be killed too. Spongy moth populations respond to weather conditions as much as anything. Wet springs trigger a fungus that controls them very well. Problems arise when you have several dry springs in a row. So if you get some good rains just before bud break in April, I would hold off on spraying. The ecological cost of spraying is very high. I don't get the sense that your populations will be high enough this year to cause serious harm to your oaks, and if the moth population does explode, you can always spray next year.

I don't understand how there can be three spongy moth caterpillars on a blueberry, two on a yellow birch, one on a redbud, a few on a basswood, and a dozen on the white oak. I have never found an egg case on these trees, despite looking. And why would a case have so few eggs in it? Do all of the larvae hatch out at the same time? Why are there different sizes of caterpillars on the small oaks at the same time?

An adult female spongy moth does not fly; she sits right where she emerged from her pupal case and releases a sex pheromone that attracts a male (males do fly). The moths mate, and the female deposits her egg mass wherever she was when she eclosed (emerged) as an adult. The next spring, the eggs hatch and the little larvae climb up as high as they can go; then they release a strand of silk and float away on the wind. This is called ballooning, and it's something that little spiderlings also do. Where they land is up to the wind. They do best when they land on an oak. Not all egg masses hatch the same day. Those on a south-facing tree trunk will warm up and hatch first. This is one reason the larvae are different sizes. Also, females are quite a bit bigger than males, even as caterpillars.

The Finger Lakes Region of New York was invaded by spongy moths in summer 2020, and as a result, many trees have hundreds of egg sacs deposited on their bark, including two old oaks on my lakeside property. Should I bring in an arborist who specializes in spraying the egg masses before the caterpillars emerge and climb upward to the canopy to devour every leaf in sight?

Spraying spongy moth egg masses with horticultural oil, before they hatch, is a good idea. It deprives the hatching caterpillars of oxygen and is deployed before leaves unfold, so they do not become contaminated. Dormant oil does not have any systemic properties, so you are not poisoning the entire tree. Ask your arborist to try to spray only the spongy moth egg masses to avoid killing other caterpillars or eggs that may be on the bark.

Where did the emerald ash borer come from?

The emerald ash borer (Coleopter: Buprestidae) is an ash specialist from Asia that was introduced to the Great Lakes Region and was first detected in 2002. Because it coevolved with Asian ash species over millions of years, the ashes have had time to developed effective defenses against this beetle. North American ashes have never encountered the emerald ash borer (EAB) and thus have no defense adaptations against it. The result has been devastating for our ashes and the 98 animal species that depend on them; since its introduction, EAB has spread rapidly south and east, killing at least 100 million ash trees so far and has caused massive problems for homeowners, land managers, and the hardwood industry. In 2007, efforts to control EAB were initiated by USDA scientists by introducing three parasitoid wasps from Asia that attack EAB eggs and larvae. Another wasp species was added to the repertoire in 2015. Fortunately, these biocontrol agents are starting to work and have reduced EAB populations and facilitated ash regeneration in parts

The emerald ash borer is an invasive beetle from Asia that has decimated our native ash trees.

of Michigan, New Jersey, and New England. Studies continue and suggest large-scale establishment and spread of these parasitoids may happen soon enough to save North American ashes.

What would you do, if anything, to protect a mature, still healthy, and gorgeous ash tree if it were under your care? We'd really hate to lose it to the emerald ash borer, and evidently there are insecticide injections that can stave off the borer. But at what cost to the rest of the ecosystem? Who besides those borers would be eating those poisoned leaves?

I won't talk you out of insecticide injections to save your tree temporarily from the borer. A dead ash tree doesn't support any caterpillars! Ninety-eight species of animals depend on ash, but we don't want to lose all of our ashes to EAB before biocontrol agents become widely established, which could still be years away in many places. I don't think you will have to treat forever, but treating now is a good approach.

Are leaves from ash trees treated for emerald ash borers toxic to litter moths and other detritivores that feed on dead matter?

I don't know that anybody has explicitly tested that with ash leaf litter, but we know that neonicotinoid insecticides (a class of insecticides derived from nicotine) can last in both soil and woody plant parts for at least 1000 days. And we know that they are systemically deposited in leaves when applied to a tree, so it is a logical conclusion that neonics would be in ash leaf litter as well. Ash leaves don't last long before they break down, so they are not a major component of litter moth diets, but you are right to be concerned.

When emerald ash borers kill many ashes in my woods, what should I do with the dead trees?

When EAB sweeps through an ash-dominated forest for the first time, there will be an overabundance of dead standing and fallen trees. Hundreds of species, including all of the species of bees that use our bee hotels, depend on dead and dying trees, along with the coarse, woody debris on the ground that is usually in short supply. I suggest you do nothing and allow the dead trees to continue to provide for these species. If some trees pose potential safety hazards, you can take them down. This might be a good time to consider purchasing a wood-burning stove!

Would you recommend cutting down ash trees that are under attack from yellow rust and inevitably from emerald ash borer? Some of the trees are up to 50 years old. We plan to replant flowering shrubs and trees, and I want to avoid dead ash trees falling on the native pollinator shrubs and trees I plan to plant. Or are we guilty of going down the same path that led to the disappearance of the American chestnut?

This is a very timely question. I don't support preemptively cutting down ashes before they get ash yellows or are attacked by borers. Neither is inevitable, and that approach would indeed be repeating the mistake we made with American chestnut. A tiny percentage of those trees may carry resistant genes to both EAB and ash yellows, and it would be an ecological shame to lose the genes that will ultimately create healthy future ash forests. But individuals that have been attacked and are obviously dying can be taken down to avoid damage to nearby plantings. Be certain your trees actually have ash yellows, though. It is a disease of ashes, but it is not everywhere.

Emerald ash borers are killing American ashes. Why don't we plant Asian ash species that are resistant to the borer?

We don't want to plant Asian ashes for the same reason we shouldn't plant any tree from Asia and expect it to fill the ecological roles that our native trees play. The most obvious ways Asian ashes will not perform like native ashes is in supporting the 98 species of animals that coevolved with North American ashes. Some may be able to use Asian ashes to some degree, but my research suggests that Asian ashes would support 68 percent fewer species than American ashes. The ultimate solution to the EAB invasion is to reduce the size of EAB populations through biological control and to encourage native ash genotypes that are at least partially resistant to EAB.

What can I do about spotted lanternfly?

Spotted lanternfly (*Lycorma delicatula*) is a recent invasive species brought to the United States as egg cases on shipments of ornamental rocks from

The spotted lanternfly is yet another invasive insect from Asia, brought to the United States as eggs on ornamental rocks.

China. We need to buy ornamental rocks from China because there are no pretty rocks in North America (*snark snark*)! Now that it's here, though, it is spreading rapidly across the mid-Atlantic States, north, south, and west. It is a homopteran with sucking mouthparts that tap into plant phloem, a part of the plant vascular system that is poorly protected from insect suckers. The good news is that there are no records of spotted lanternfly killing trees, even when they appear in large numbers on the trunks and limbs. The bad news is that they love grape vines and apple trees and can ruin fruit production in these industries.

As nymphs, these insects develop best on ailanthus, the tree of heaven (or is that hell?), an invasive, smelly tree first brought to the eastern United States in 1784. So step one in controlling the lanternfly is to make sure you remove all ailanthus trees from your property. There is a two-fold benefit to getting rid of these trees: not only will this reduce the rate at which lanternfly populations grow, but it also removes the host plant that, when eaten, makes these insects taste bad to predators. Spotted lanternfly is brightly colored red and black, a standard warning (aposematic) coloration in the animal world that signifies they taste bad. But fledgling birds near your house have not yet learned that red-and-black insects taste bad, so when they try a spotted lanternfly that *hasn't* been hosted on ailanthus, the birds will find that they actually taste good. Over the last few years, I have watched this happen at my property, which has a steady population of lanternflies. I suspect the birds on my property have something to do with keeping the population in control, because I regularly see the insects' wings on the ground. Unless you own a vineyard, I recommend avoiding insecticides to control spotted lanternfly, especially the highly toxic neonicotinoids that are too commonly used as systemics—they last for years; poison your plants against *all* insects, not just the bad guys; and are unnecessary for spotted lanternfly.

Are spotted lanternflies poisonous? I'd like to feed them to my bluegill, sunfish, and largemouth bass. I don't want to learn the hard way, and I find conflicting data online. I'm hoping you can offer your best advice?

There is conflicting data online because their edibility depends on what they have been eating. If spotted lanternflies eat ailanthus or milkweed, they will taste bad. Your fish would probably spit them out before they caused harm. But if they grew up on grapes, maple, Virginia creeper, or other natives that are not loaded with toxins, they will not be distasteful.

I haven't seen any spotted lanternflies yet. Do you have an opinion on how to trap them?

They are coming your way, and you can go to great lengths to trap them, but you won't stop the invasion. The adults are adept fliers and can move a good distance in late summer and fall. If you have ailanthus on your property, which spotted lanternflies love, this is a good time to get rid of it. If you like fiddling with things, you can find information online about fashioning a trap for nymphs. Hosing nymphs off a tree trunk with soapy water works too.

Do spotted lanternflies hurt oaks?

Spotted lanternflies will feed on oaks as adults, but oaks are not their favorite hosts. There are no records of them killing or even hurting oaks.

CHAPTER SIX

Pest
Control

How important are the moths of hornworm caterpillars as pollinators?
We all know people hate these caterpillars because they eat tomato
plants. What should we do?

Hornworms transform into sphinx moths, which are valuable pollinators for
flowers with deep corollas (blue lobelia comes to mind). But perhaps more
important is the caterpillars' role in the food web. Sphinx moth caterpillars
form a major component of nestling bird diets, particularly cardinals. The
best control for tomato and tobacco hornworm are tiny braconid wasps. Those
white "eggs" that you see on the back of a hornworm are actually cocoons of
this parasitoid that have already consumed the innards of the caterpillar. If
you squish a hornworm loaded with those braconids, you will kill dozens of
the natural enemies that keep hornworms in check. So the trick is to ensure
that you have braconid wasps around when you need them. Paradoxically,
the way to control hornworms is to encourage their presence and thus the
presence of their natural enemies. You can do that by including plants in your
yard that host other species of hornworms, such as black cherry, ash, Virginia
creeper, American elm, and fringe tree. I have 17 species of sphinx moths in
my yard because I have planted their larval host plants. When tobacco hornworms emerge, the braconid wasps arrive, ready to attack them.

The tobacco hornworm has many natural enemies, particularly tiny braconid wasps that develop inside the caterpillar and then pupate on its back.

What are your thoughts about the value of releasing beneficial insects, such as minute pirate bugs and *Neoseiulus* species predatory mites, to control pests in a small home garden with a history of some pesticide use?

In theory, adding predators to a garden can suppress pest populations, but it is difficult to make this work. Large pest populations must be present when you release the predators so that the predators can increase their own populations. The fact that pesticides were once used in the garden shouldn't be a problem unless they depressed the pest populations below what is needed to keep the predators around. If there are not enough pests for the predators to eat, they won't stick around. Also, minute pirate bugs and predatory mites are common as long as they haven't been wiped out by pesticides, so you don't need to bring them in. Predators are always the first insects to die when you spray. The hard part for most gardeners is to be patient enough to wait for predators to find the pests and then grow to adequate numbers to control them. That requires some tolerance for pests, which most gardeners are

When periodical cicadas lay eggs in branch tips, the branch often dies, causing flagging.

taught to avoid. In short, I wouldn't bother to introduce predators. I would instead try to create conditions that attract predators naturally—and that, of course, means no insecticides.

I've noticed dead branch tips on everything from oaks to dogwoods and viburnums, along with evenly spaced "tears" along the bark? Is this a new insect?

This is known as flagging, and it's caused by native periodical cicadas. After females insert their eggs into the plant's thin branches, the branches often die. I call them "nature's pruners." The plants will be fine next year, but they won't grow much this year. Periodical cicadas appear as adults only once every 13 or 17 years, depending on the species and brood, so flagging does not happen often enough to be particularly harmful to trees and shrubs.

We have a live oak in our yard, and just about every year, buck moths show up. I have heard their sting is a bad one. I have yet to be stung. Are there any known predators for this caterpillar?

Several bird species will eat buck moths, but these spiny caterpillars are not the birds' first food choice. I once watched a Prothonotary Warbler feed one to its nestling, but not before whacking the caterpillar silly against a branch to kill it and dislodge the spines. That worked, but it probably took more time and energy than most well-fed birds care to invest. Thanks for not spraying!

How should I control Japanese beetles?

Japanese beetles were introduced to the East Coast on nursery stock in the early 1900s. Related Asiatic beetles that behave similarly to Japanese beetles were introduced somewhat later. As adults, these beetles are generalists, although they have an affinity for roses, sycamores, sassafras, evening primrose, and Virginia creeper. But as larvae, they are grass specialists. One reason they have been able to spread so successfully is because there are 44 million acres of lawns in the lower 48 states that serve up the larvae's favorite food: grass roots. The obvious way to decimate the beetle is to reduce its primary resource, and many of us in the United States are doing just that. Across the nation, people are reducing the area of their lawns.

As we reduce the size of our lawns, we can also apply a naturally occurring bacterium, milky spore disease, which kills Japanese beetle larvae. The fungi can last up to 10 years, but it works very slowly and does not spread far from where it was applied. You can also apply solutions containing two species of nematodes that attack both Japanese and Asiatic beetles, *Steinernema glaseri* and *Heterorhabditis bacteriophora*, both of which are available commercially. Or you can buy Japanese beetle traps baited with sex pheromones that attract beetles from good distances. These traps can kill a lot of beetles, but you have to empty them as soon as they get full. Otherwise, you will attract beetles to your yard that won't be able to fit into the trap! When my property has an outbreak of Japanese beetles, I fill a pail with hot, soapy water and then walk around the yard and hold the pail under leaves that have lots of beetles. Even a slight bump will knock the beetles off the leaves into my pail, where they die quickly. It's surprising how effective this simple method can be, especially if you start early in the season as soon as you see the first beetles (around July 4th at our house in southeast Pennsylvania).

Do sawflies have any redeeming value? I hate them!

Yes, they do, at least if you are a hungry bird. Sawfly larvae look an awful lot like Lepidoptera caterpillars—so much so that birds do not distinguish between them when they are feeding their young. Remember that it takes thousands of caterpillars to feed one nest of baby birds, and the presence of sawfly larvae in your yard can easily spell the difference between nesting success and failure for local birds. Years ago, a bluebird in my yard reared her first brood almost exclusively on sawfly larvae she found on our white pine trees. It had been a cool, wet spring and most of the moth caterpillars she normally would have relied on weren't around that year. Good thing for the bluebird family that sawflies were!

This is a troubling time of year for seeing fall webworms. I'm OK with them now, but people are upset because their webs make the trees look awful. I don't find a great deal on the internet in their defense. Don't they turn into valuable pollinating moths?

Webworms are troubling only because we have decided that only pretty creatures are allowed to share our landscapes. Webworms make ugly webs, but they rarely harm the trees in any measurable way. They are one of the cogs in the great wheel of biodiversity, and they turn into moths that help feed bats, nighthawks, and whip-poor-wills. The moths don't pollinate to my knowledge, but pollination is only one of the benefits that insects bring us. Another huge benefit is that they supply energy to the food web. Webworms eat leaves and then, as adult moths, become food for many creatures that do not eat leaves. That's how energy moves from plant to vertebrate.

Webworms and tent caterpillars, the insects that make white silk tents in the branch crotches of cherries and apples in spring, are both eruptive species, and their populations explode (erupt) in some years, while in other years you can't find any. Last year I counted 100 fall webworm tents on my property. A

Yellow-billed Cuckoo (a specialist on hairy caterpillars) came along, sliced open every one of them, and ate all of the caterpillars. This year I don't have a single tent—but I know they will be back. If you can't stand their look, and if you don't have any cuckoos, get a stick and slice open the tent. That will give parasitoids and other insect predators access to the caterpillars inside. Even other birds such as Baltimore Orioles will take advantage.

I've never used any type of chemical in my yard at all. Recently, however, all the termite companies are insisting on using termite baits, which are to be buried in the yard. What do you think of these? Do these affect the other insects or disrupt the environment?

Worry not. Termite baits are a huge step forward in protecting our homes from termite infestations. The baits are basically pieces of wood doused with insecticide. After termites find the wood, they take the material back to their nests. This is the most benign way to treat termites that I can think of.

Is there a way to avoid harming stinging insects that need to be removed from municipal parks and street trees? I work for a locality and we are constantly spraying paper wasp nests.

I applaud your kind sentiments toward the natural world, even when it stings. It *is* possible to move the nests of social wasps, but only early in the season when the nest is small with few workers that are bent on defending it. But this is a risky adventure, with low success rates in reestablishing a nest somewhere else.

I am considering buying a house that has had a termite infestation, carpenter ants, and bees. Every pest control expert wants to use something like fipronil. This horrifies me, because it might poison both the well water and the land—should I take their advice?

There is a clear conflict of interest if you get your advice from a pest control person. The only wood-eating insects you need to be concerned about are termites. Treating your property with insecticides just in case you have termites is not the way to go. You should treat only if you *do* have them, and you should do it through bait stations that attract only termites. Carpenter ants do not eat wood—they can hollow out and live within wood that has already been compromised by dry rot, but they do not themselves destroy wooden structures. The nests of the carpenter ants you occasionally see in your yard are usually in trees. Carpenter ants forage inside your house for spilled food, and usually only in spring. This can be a nuisance, but they are harmless if you ignore them. If you do have a carpenter ant nest in your house, you need to replace the dry-rotted wood in which they have made that nest. Carpenter bees nest in untreated pine and redwood, but it would take many years to structurally compromise a piece of wood. The best way to keep them out is to paint the wood. You might also give them a nesting alternative: a horizontal pine board for them to tunnel into instead of your house.

Do you approve of systemic insecticides like neem oil or rosemary oil? What do we do about boxwood leafminer, crape myrtle scale, tea scale, and on and on—pests that we really have no way to combat except with a systemic? And one last question: We do invasive plant removal, and the only way to keep tree of heaven, privet, and others, from resprouting is to paint with a systemic herbicide. We hate it, but our work is in vain if we don't paint; plants just grow back. What do you say to this?

First, we have to be clear on terms. A systemic insecticide is carried to all parts of the plant, so when an insect feeds on the plant, the insect dies. Most systemics, including neem oil, are indeed transported to pollen and nectar as well as leaves, so flower visitors are exposed to them. I don't know of any research that has actually tested the effects of rosemary oil. My guess is that it's not very strong and probably won't rid your plant of pests. As for systemic insecticides on non-natives like boxwood and crape myrtle, almost no native insects use those plants, so you won't be killing insects that are a valuable part of the food web by using a systemic on Asian ornamentals. An herbicide, in contrast, is designed to kill the *plant*, not insects that eat the plant. I use herbicides (stump paints) on woody invasives when I can't whack them out with a mattock.

As a conservationist and entomologist, can you recommend a way to control the common occurrence of ants coming into my house to find moisture, sweets, or proteins? They persist in finding ways in.

I'm afraid I will be a big disappointment on this one. Ants that move into houses are notoriously difficult to control. Their colonies are typically located outside of the house (under your roof shingles, in trees in your yard, and in other places), and they are only foraging in your house. Ant baits that ants carry back to the nest are commercially available, and they work to some degree. Ant foraging is seasonal, and at our house we just put up with it and wait them out. I would not recommend any kind of toxic spray. You will be living in an envelope of insecticide with unknown, but likely bad and long-term, health effects on you and your family.

I have noticed in the past few years that my crabapple and hawthorn trees start out great in spring, but by August they are looking bad with very early leaf loss. I have seen this not only at my house but at other houses too. Any ideas what could be causing this anomaly?

Sounds like your trees have apple scab, a fungal disease caused by *Venturia inaequalis*. The best way to avoid it is to plant one of many resistant cultivars. But if you're going to switch out your Asian crabapple for a resistant cultivar, consider one of our native crab apples, such as *Malus coronaria*, *M. angustifolia*, or *M. ioensis*.

Waxy, white leaf eaters are eating up the leaves of my red twig dogwood. What do you suggest I do, if anything? Should I assume the tree will survive? Will these turn into something beneficial in the food web?

Not everything is equally beneficial, and you're describing the dogwood saw fly, which does little for the food web as a caterpillar because it covers itself with nasty-tasting wax. As of now, it is winning the evolutionary arms race between predator and prey. So while we are waiting for predators and parasitoids to figure out how to control its populations, feel free to pick them off your dogwood or knock them into a bucket of soapy water. They can defoliate your dogwoods, which seems awful but doesn't actually harm them. I have gray

The dogwood sawfly covers itself with wax as a very effective means of warding off predators.

dogwoods that have been defoliated for 10 straight years, with no ill effects to the shrubs. They are back smiling the next year.

What eats slugs?

Native slugs serve as food for box turtles, firefly larvae, nematodes, carabid ground beetles, ducks, Downy Woodpeckers, robins, grackles, and other creatures. Unfortunately, most of the slugs we encounter in our gardens are introduced species with fewer natural enemies.

How can I control cabbage white butterflies?

Cabbage white butterflies are invasive insects from Europe that attack crucifers of all kinds, such as cabbage, kale, broccoli, arugula, Brussels sprouts, and bok choy. They are not native, are not part of important food webs, and are here without any natural enemies to control them. Therefore, controlling them in our gardens has no untoward ecological ramifications. Certain formulations of *Bacillus thuringiensis* (BT), a natural bacterium, target caterpillars of all kinds, so I always hesitate before recommending its use in natural areas. But in our vegetable gardens, BT is an excellent, safe alternative to other insecticides and usually gives good control of cabbage whites. Note, however, that if you are enjoying black swallowtail caterpillars on your parsley, dill, or carrots, BT will kill them as well. So be careful where you spray. Fun fact: Shine an ultraviolet light on the pure white wings of a cabbage white to experience how cabbage whites see their fellow species. If it's a female, the wings glow lavender; if it's a male, they shine brilliant royal purple.

What are your thoughts about beekeeping near golf courses? This is becoming a trend for golf courses, and I'm wondering if there is a risk of harming the bees if the course is using pesticides? I'm also wondering if the honey produced might contain pesticides?

The big threat to honey bees (and all insects) today is neonicotinoid insecticides. They are systemic and, when applied, are translocated to all parts of a plant, including the pollen and nectar. Whether or not the beehives are located on the edge of the golf course property, their bees would be able to reach all of the plants on the entire golf course. So the question is, does the golf course use neonicotinoids or not? Whether neonics accumulate in the honey produced by bees in these hives, I can't say. But neonics *do* harm insects, and on a golf course, they are probably not even necessary. If the course does not use neonics, the bees are in the clear; if neonics are used, please urge them to stop!

Will cutting back my milkweed control aphids?

Yes, that is one of the best ways to control the oleander aphid, an invasive species from the Mediterranean region. Another is to literally (and gently) wash your plant with soapy water.

How can I kill carpenter ants in my trees without hurting caterpillars?

Commercially available carpenter ant baits will do the trick, but I have to ask why you want to kill carpenter ants when they are behaving themselves in your trees? They are useful predators that help keep your ecosystem in balance. They are also used by Pileated Woodpeckers and flickers to rear their young. Please don't deprive these beautiful birds!

Oleander aphids, pests of milkweed, were introduced from Europe, as was its primary host, oleander.

What do you think of dormant oil?

Dormant oil is good for smothering scale and hemlock woolly adelgid, the tiny, aphid-like creature that is killing hemlocks across eastern North America. It also works for smothering spongy moth eggs. I like it for those uses because it kills by cutting off the insect's oxygen, not by injecting a systemic insecticide into the plant that will poison the plant for years. But dormant oil also kills the caterpillars of moths and butterflies, as well as their eggs and pupae. So I would use it only when absolutely necessary—and if your hemlocks are infested with woolly adelgids, it *is* absolutely necessary!

Is it normal for tulip poplars to exude sap from aphids?

All aphids produce honeydew, a sugary byproduct of the plant sap they consume. Plant sap contains very few nutrients, especially nitrogen, so aphids and other homopterans must suck up enormous quantities of sap to extract all the nitrogen they need. Natural selection has devised a clever way for aphids to handle all of that sap: a portion of their digestive system has been modified into a filter chamber, a structure that extracts the minute amounts of nitrogen in the sap before sending it through the aphid's intestinal tract. Once the nitrogen is removed, the sap is shunted directly out the aphid's anus in the form of sugar water we call honeydew.

Tulip trees, also known as tulip poplars (*Liriodendron tulipifera*), support a specialist aphid (*Illinoia liriodendri*) that becomes abundant once in a while. When aphid populations build up on the trees, the insects produce a lot of honeydew, which accumulates on the leaves and forms the substrate for sooty mold, a black fungus. In years with particularly large aphid populations, black sooty mold can cover most of the leaves, trunk, and ground beneath tulip trees. It's unsightly, but we're wise to tolerate it, because the aphid colony will inevitably be attacked by ladybird beetles, predators that will devour every last aphid. If you spray the aphids with an insecticide, you will kill the ladybird beetles and prolong the aphid problem. Tulip tree aphids rarely explode two years in a row. So is it normal for tulip trees to become covered with honeydew? It is, but only every now and then.

Are bug zappers hurting insect populations?

In 1995, while my wife and I were waiting in line at a toll booth on the New Jersey Turnpike, I wondered how many of the insects being drawn into the bug zapper hanging from the booth were actually the zapper's targets—biting flies, such as mosquitoes, ceratopogonid midges, stable flies, horseflies, and deer flies. Later, Tim Frick, a local high school student who was looking for a research project, agreed to help me figure it out. He put trays beneath bug zappers in several yards in Newark, Delaware, and collected the insects killed by the traps once a week throughout that summer. Then I identified the bodies. The results surprised even me. Only 0.02 percent of the dead insects were actually biting flies; 99.98 percent of the insects killed by those traps were harmless non-target species such as chironomid midges, moths, and caddisflies. The zappers were clobbering populations of nocturnal insects.

Fortunately, that study helped reduce sales of bug zappers, but they are still in use today in some circles, and they still don't work as advertised. In fact, a new silent insect trap called DynaTrap is being sold these days and promises to catch mosquitoes but not honey bees or other pollinators. I had the opportunity to look at three weeks' worth of insect catches from one of these traps. It contained more than 10,000 insect bodies that included crane flies, fungus gnats, midges, many small moths, and assorted other insects—but *no* mosquitoes. The woman who owns the trap insisted she had lots of mosquitoes in her yard, and I believe her—but the trap has not been catching them. Yes, bug zappers and more modern insect trap designs hurt insect populations—without controlling mosquitoes!

I live in northern Virginia, and we often have outbreaks of fall canker-worm. The response has been to spray wide areas, even those not having a problem with cankerworm, with BT to "control" the cankerworm. Is there a better way to handle this problem?

My first inclination is to make an unpopular suggestion: the best way to handle a fall cankerworm outbreak is to do nothing, be patient, and allow the insect's natural enemies to bring them under control. No one was spraying trees before we invented insecticides, yet all of the trees didn't die. Case-by-case spot spraying of individual prized trees with BT, rather than blanket spraying thousands of acres from helicopters or trucks, is the next best option. Spraying for cankerworm, an inchworm species that birds love to eat, kills the cankerworms, but it also kills all other caterpillar species. And when you reduce caterpillar populations, you reduce the populations of the many species of natural enemies that buffer cankerworm outbreaks. Under natural conditions, insect predators, parasitoids, and diseases track canker-worm outbreaks, eventually bringing them under control. Spraying delays or prevents these natural controls, but a "one-time" spraying event often leads to repeated, expensive resprays that only prolong the outbreak. Past sprays in Fairfax County, Virginia, provide a perfect example. A "one-time" spray in the year 2000 led to repeated sprays in 2001, 2002, and 2003. Similarly, a "one-time" spray in 2012 led to additional sprays in 2013, 2014, and 2015.

There is a serious ecological downside to reducing spring caterpillar populations. Springtime caterpillars are the primary fuel for 39 species of birds that migrate through Virginia, as well as for 65 species of resident birds. Flocks of migrants eat tens of thousands of caterpillars when they settle in an area during refueling stopover events. But migrants must refuel quickly, and if caterpillars are scarce or absent altogether, as they are after a spraying event, the migrants move on. Each year, there are fewer and fewer places for the birds to move to, and migrant populations are declining rapidly as a result.

Rapid declines in both migrant and resident North American birds are now being recorded by formal studies. A 2019 study by Kenneth Rosenberg and other researchers revealed that we have lost three billion breeding birds in North America—about one-third of all North American birds—in the last 45 years. Because birds are such important predators of caterpillars, bird declines create a negative feedback loop that encourages cankerworm outbreaks. As we lose more and more birds, the predation that once limited cankerworm outbreaks is lost, and if we spray cankerworms during outbreaks, we lose breeding and migrating birds.

I dug up a very large larva that was some kind of bad borer. Would it have been OK to put it out for the birds?

How do you know it was a "bad" borer? If you dug it up, I suspect it was the larva of one of the June beetles, probably green June beetle. Nothing bad about them. They eat grass roots, and if you are like almost all of us, you have too many grass roots in your yard.

Can you put it out for the birds? Yes, and they will love it—particularly grackles and flickers that poke their bills into lawns to find those grubs. But please think about your knee-jerk assumption that the larva was "bad." We humans live off the interest produced by our "ecological bank accounts." We call that interest "ecosystem services," and it is our life-support system. Because of their importance in running our ecosystems, insects are the currency in our ecological bank accounts. Just as it would be unwise to call a dollar bill a "bad" dollar, it is unwise to assume that every insect you find is "bad."

I live in the suburbs, and some of us who garden for wildlife are being cited by the town health department for creating "rodent habitat." Both of my neighbors have had mice in their homes, and I had them nesting in my garage last winter as well. What would you recommend for a situation like this? I have a screech owl box, but I am concerned that rodenticides will reach the owls through the food web, especially being adjacent to a high school that has a pest-control regimen.

June beetle larvae, often called grubs, are a favorite food for many birds.

Mice in the house are a pain in the neck (something my wife and I know well), but mice in the landscape are a necessity. A landscape without mice is nearly as dead as a landscape without insects, because mice are an important part of the local food web. So the question is not how do we remove mice from our landscapes? But rather, how do we keep them out of our houses? I won't sugar coat this: it's difficult! Mice are excellent at entering our houses through the smallest holes, and I wouldn't recommend spending a lot of time trying to make your house so tight that they can't get in.

Before I go on, I need to distinguish between mice and rats. The Norway rat is an invasive species from Europe and is not a crucial component of local food webs. These rats live with people and are not attracted by well-planted yards. They are attracted to our garbage, our bird seed, and other human stuff. If you have rats, you have to kill them. I suggest you use rat traps; when these traps are used properly, they can catch rats in a single night. Please do not use rat poison, though. It will kill a rat, but often not before the rat goes outside in a stupor and becomes easy prey for a local owl, hawk, or fox. Then, of course, your rat poison kills a lot more than rats. For example, studies have shown that rodenticide use in coastal South Carolina has decimated the bobcat populations.

Back to mice. Pest control companies will tell you that mice spread several deadly diseases, especially hantavirus. This is hyperbole. It is true that a heavy mouse infestation can harbor hantavirus, but it is extremely rare and mostly confined to drier states in the West. Between 1993 and 2017, 697 cases of hantavirus were reported in the United States, with fewer than 10 fatalities—that's 24 cases per year. Just to put that in perspective, during the same period of time, 76,000 people were hospitalized by injuries caused by cell phones, with 16,000 related deaths between 2001 and 2007. Moreover, although white-footed mice do carry hantavirus and they can get in your house, the most common house resident is the house mouse, a species native to India but long associated with humans. This is the same cute little mouse sold as a pet at your local pet store. The good news is that house mice do *not* carry hantavirus. We should keep the danger aspect of mouse infestations in perspective.

Mice move into our houses when there is more food inside than outside, typically during winter months. Open dishes of cat or dog food are a bonanza for mice, but sometimes just a warm, dry refuge is enough to lure the little guys inside. I know most people assume that if you have a well-planted native landscape, you will have more mice in your house than if you have a traditional landscape, but I don't know of any studies that have tested that assumption. I can offer one data point from my own experience that refutes it: we lived in a typical suburban development for 10 years before we moved to and created our current native landscape. We had just as many mouse issues in the old house as we do now.

What to do? There are two good options that don't depend on poison. The first is to live trap your mice and then take them far away. And I mean far away—50 yards won't do it. In our old house, I once caught a mouse in a live trap and carried it at least 75 yards away in the backyard. The mouse ran back to the house faster than I did. I kid you not! I call the second method of mouse control the "no more Mr. Nice Guy option." Old-fashioned snap traps work well and can be purchased at your local hardware store.

I try to be as ecologically minded as possible, but I have several enemies that pupate (I guess?) in fall leaf litter: azalea sawfly and lily leaf beetle. I'm concerned that if I don't do a good fall cleanup, it will encourage their proliferation, but I realize this might sacrifice beneficials. What should I do?

Compromise is not a dirty word. If you are landscaping ecologically on most of your property, cleaning up near your azaleas and lilies if you have pest problems is OK. By the way, lily leaf beetles are not native, so you certainly don't want to encourage them. I question the degree to which your leaf litter is giving them refuge, though, because both lily leaf beetles and azalea sawfly larvae pupate in the soil, not the leaf litter. Raking away all the leaves near your plants will not get rid of these insects, but it will hurt the soil mycorrhizae your plants depend on.

I have a question about rust, especially in regard to amelanchier trees, hawthorns, and so on. Where I live, cedar-quince rust is a big issue because of the Callery pears as well as cedar-apple rust from neglected home "orchards." What is the best way to deal with this issue? Being in the suburbs, it is hard to remove all alternate hosts because your neighbor is likely to have a juniper even if you don't. Neem oil would affect insects, and I read that copper may be affecting bees. Have you had this issue, and are you aware of any remedies that can be used without sacrificing the functionality of the native planting?

We have the same conflict between eastern red cedar and amelanchier, American plum and apple trees. Our pathogen is cedar-apple rust, a fungus related to cedar-quince rust. If I were growing those trees solely for fruit, I might think about removing some of my cedars. But cedars are great trees for cover, caterpillars, and fall berries. So my compromise is to do nothing. I have cedar-apple rust and don't get viable fruits from my amelanchier or fruit trees. But otherwise, the trees are healthy and produce lots of early season flowers for our native bees and lots of good foliage for our caterpillars. You could use myclobutanil, if you are up for going that route. It's included in several commercial products. A new biofungicide,

Fungal growths from cedar-apple rust appear in spring on eastern red cedar.

Bacillus subtilis, is also available in commercial products and is supposed to be nontoxic to bees.

Bagworms are native, so I assume birds eat them. Do you suggest clipping off the bag or leaving it there, considering the permanent damage the bagworms make?

There are several species of native bagworms, but the one that is most common and causes the most problems is the evergreen bagworm. Cutting off the bags with scissors does work, particularly if you do it during winter. At that time, the bag is either empty because it once contained a male that has already emerged or it contains a female that is stuffed with eggs that will hatch in spring. Getting rid of all of the female bags gets rid of the next generation of bagworms.

Bagworm problems are a sure sign of a landscape without enough plant diversity. Gardens typically get a lot of bagworms when they include too many arborvitae conifers that bagworms love. Diversify your hedges, and bagworm populations won't build up. If you spray for these "pests" and mosquitoes,

Birds *do* eat bagworms, but so do many hymenopteran parasitoids.

the spray will also kill the many parasitoids that attack bagworms. On our property, we have many eastern red cedars, another plant bagworms love, but I have to search to find a bagworm. Lots of predators, from chickadees to ichneumon wasps, keep them in check.

My swamp white oak was being attacked by some pest, and our tree people suggested a yearly spring application of a soil pesticide. Without knowing more, can you say whether this is needed?

I suspect your tree people are taking advantage of you. Being "attacked by some pest"? What "pest"? What are the symptoms? Lots of insects develop on oaks, but none really hurt the trees other than spongy moth, an invasive species from Europe. If you have a spongy moth problem, so will everyone else around you, and you won't have to wonder what is hurting your oaks. And if you *do* have spongy moth, a soil insecticide treatment won't help at all because they do not pupate in the soil. I suspect that if your oak is actually suffering, it is from one of the introduced oak diseases. Soil pesticide applications won't help that either. Permit me a generality, if you will: never, ever treat by the calendar. When you do that, you are applying deadly insecticides whether or not you need them. The misuse and overuse of insecticides is one of the major causes of global insect declines—something that threatens ecosystems everywhere.

Mosquito Control

Do mosquitoes serve any ecological function?

Yes, indeed, they do! Mosquito larvae are an essential node in the food web of freshwater ecosystems. Most of the species that live in aquatic systems are mosquito predators, including dytiscid predaceous diving beetles, nepid water scorpions, notonectid backswimmers, belostomatid giant waterbugs, zygopteran damselflies, and anisopteran dragonflies. So are frogs and toads, immature newts and salamanders, and fish of all stripes. They all rely on mosquito larvae for food. And once mosquitoes emerge as adults, they support adult dragonflies and damselflies, swallows, martins, swifts, and bats. Eliminating mosquitoes would impact many hundreds of aquatic and terrestrial species. But several troublesome mosquitoes, such as *Aedes aegypti* (the yellow-fever mosquito) and *A. albopictus* (the Asian tiger mosquito), are invasive species. They were never part of North American ecosystems, so waging war on them will not have ecological ramifications—unless we also kill everything around them.

I have noticed a growing number of neighbors signing on to mosquito- and tick-control services. Not coincidentally, my garden has become quieter—fewer crickets and other insects at night—and seems to have fewer pollinators. These "pest control" services advertise organic treat-

ments to reduce and repel ticks and mosquitoes. The literature focuses on the ability of essential oils to repel insects, but my hunch is that the application of these oils, as they come in contact with any insect, is disruptive and likely lethal. Do you have any knowledge of the impact of these essential oils and other "organic" treatments more broadly on insects? My neighbors are choosing "organic" mosquito treatments because they believe they are safe for the environment.

I don't know of any studies that have examined non-target kills by essential oil products—whatever they are—for controlling ticks and mosquitoes. But let's look at this logically. *Organic* means nothing in terms of toxicity. Cyanide is 100-percent organic and 100-percent deadly. So is ricin. Think about it: Could a group of essential oils kill mosquitoes and ticks (completely unrelated arthropods) without killing all the other arthropods it comes in contact with? Including the pollinators and caterpillars so vital to local food webs?

What is the best way to control mosquitoes in your yard?

The most effective and benign way to control mosquitoes on a small scale is using biological control with *Bacillus thuringiensis* (BT) formulated for aquatic Diptera in mosquito dunks. This inexpensive product is available from your local hardware store. During mosquito season, typically June through October in most places in the United States, fill a bucket (at least two-gallon, black bucket) with water and add a handful of organic matter—straw, hay, or dead grass clippings all work. Then put the bucket in the sun so that good populations of algae and diatoms build up in the water. That's what mosquito larvae eat, so your concoction will be very attractive to female mosquitoes looking for places to lay their eggs. After four or five days, look for little wrigglers (mosquito larvae) in your bucket. When you see them, add a mosquito dunk. The larvae will nibble on it and die. A mosquito dunk is very targeted and won't hurt anything but aquatic dipterans, and the only aquatic dipterans in your bucket will be mosquito larvae. If your dog drinks from the bucket, it will be fine. If a dragonfly lays eggs in your bucket, the dunk won't hurt the resulting nymphs. Put a coarse screen over the bucket so chipmunks or mice don't drown. That's all there is to it.

When should I start a mosquito dunk bucket? Must it be always above freezing?

Start your bucket in early summer when mosquitoes start to become annoying. No point having a mosquito dunk bucket when it's too cold for mosquitoes to be actively breeding. Depending on where you live, I would say dunk buckets are best from June through October.

How long does it take the hay/straw mixture to produce the algae and diatoms that mosquito larvae eat?

This should take three or four days if it's warm out. But it may take another four or five days until you actually have mosquito larvae swimming around. Then you can drop the dunk in. Mosquito larvae are not invisible; you can easily see them wriggling around in the bucket.

Will female mosquitoes continue to lay eggs after the mosquito dunk is added, or does adding the dunk destroy the bucket/straw attractiveness? What's the attractant range of the bucket/straw setup? Will multiple buckets around a property work better than a single one? If female mosquitoes live seven to ten days and have a range of one to three miles, are mosquito dunks good for eliminating/lessening mosquitoes for an average quarter-acre lot, or is the bucket continually attracting new mosquitoes?

You would be surprised to learn how little research has been done on mosquito dunks. But here are my best guesses. The dunk will not reduce the attractiveness of the brew. The attractant range of your bucket will depend on the size of the bucket. A big bucket with lots of brew should easily attract all the female mosquitoes near a typical suburban yard. But multiple buckets will cover a larger area and will have a better chance of attracting more mosquitoes. Remember that your bucket is competing with surrounding water sources. If there are no other water sources nearby, the bucket will be the only game in town for the female mosquitoes, and they will find it. Although a mosquito can travel up to three miles in its lifetime, especially with a wind event, most mosquitoes will rarely travel more than 100 yards from where they emerged if there is no need to do so. Will your bucket attract mosquitoes to your yard? Maybe—but only gravid (pregnant) females looking to lay eggs. Host-seeking females looking for a blood meal are attracted by carbon dioxide and octanol, a component of cow breath—not larval habitat. Mosquito dunk chunks or bits should be just as effective as the larger dunks, although they are not as long-lasting.

Why is a mosquito dunk needed? Couldn't the water simply be poured out after a few days to kill the larvae?

Yes, you could just dump the bucket periodically, and if your timing is good, that method would work. But then you have to set it up again. And if you forget to dump it, you have produced a whole bucket of mosquitoes.

How can you tell if the mosquitoes are laying eggs in a trap—are the eggs visible? I found a web page that says, "The eggs are so small you can barely see them without a magnifying glass," so I'm curious how one would know whether the mosquitoes have laid their eggs?

Many mosquitoes lay rafts of eggs that float on the surface of the water. They are easy to see. But the larvae are even easier to see once they hatch. They are always wriggling around in the water. You just have to add your dunk before the larvae mature, and you have a five- or six-day window. I would wait until you see larvae wriggling around before adding the dunk.

Can honey bees drink from a mosquito dunk bucket?

Yes. Remember that mosquito dunks kill only aquatic dipterans. It is harmless to insects in the Hymenoptera, which includes bees, ants, and wasps.

How can I address my neighbors who are using fogging companies to control mosquitoes? I am trying to create a garden with more native plants, and I am worried that the spray on their property is going to affect the insects on my property. Is that a valid concern?

Unfortunately, that *is* a valid concern. Mosquito fogging kills every insect the fog comes in contact with. And the fog inevitably drifts from your neighbor's yard to your yard. There is not much good to say about mosquito fogging.

But how do you convince your neighbors that they are doing far more harm than good by hiring a fogger? Your first approach might be to make sure they have the facts. Mosquito foggers often claim that their fog kills only mosquitoes. That's not so. As I said, it kills all kinds of insects, including the beloved monarch butterfly, fireflies, and all of those pollinators we are trying to protect. Fogging companies also say that, because the fog is formulated from pyrethrin, a natural product produced by chrysanthemums (they don't mention that it is industrial strength pyrethrin), it is harmless. But harmless to everything except mosquitoes? Does that make sense?

Although fogging does kill mosquitoes, it never seems to reach enough adults to control the population. Mosquitoes are notoriously difficult to control in the adult stage. You have to kill 90 percent of the population to get good control. Fogging kills between 10 and 50 percent of the adult mosquito population—not even close to the amount required to get control. The sad fact is your neighbors are spending a lot of money for a service that does not work as advertised. But if you offer a more targeted, cheaper alternative that actually does work—mosquito dunks—they might listen to you.

These monarchs, along with thousands of others, were killed by a single mosquito fogging event on Kent Island, Maryland.

I had a thriving ecosystem, which Sarasota County, Florida, mosquito control has now destroyed. I have an exemption from mosquito spraying and yet my property has been sprayed multiple times, and now all of our caterpillars and butterflies are dead, including the endangered Atala. I have written my county commissioners and talked with mosquito control, but I haven't made any progress. What else can I do?

I'm very sorry to hear this. I am not a litigious guy. In fact, I have never sued anyone and doubt I ever will. But I think there is a role for lawsuits when the big guy persists on rolling over the little guys just because they are sure they can get away with it. Perhaps a well-worded letter from your lawyer to the county commissioners will get their attention. They simply do not have the right to kill everything in your yard. Maybe members of the Florida Native Plant Society will join you.

If I have fish in my pond, will I have mosquitoes?

No, you won't, and you are not likely to have mosquitoes even if you don't have fish in your pond. Mosquitoes do poorly in permanent bodies of water, because nearly everything that lives in those waters eats them, fish being only one of those predators. Temporary bodies of water, such as the pools in poorly drained rain gardens, ruts, tires, clogged gutters, and other places—that is, waterbodies with no predators—produce most mosquitoes.

Deer Control

How does one control deer? We all complain, but I don't see anyone doing anything about it.

There is no scientific mystery awaiting discovery that is keeping us from controlling deer. We know how to control deer. In fact, deer were controlled so well in the past that they went extinct in Pennsylvania and other parts of the eastern United States around 1900 and had to be reintroduced from herds in the West. All populations are controlled by one of two forces: top-down pressure from predators and diseases or bottom-up control by the loss of resources, usually food. Let me stress that no population (including humans)

can grow forever without running up against insurmountable ecological bar-
riers. In the case of deer, we have removed top-down control from predators
(although not from chronic wasting disease, which is rearing its ugly head
in many regions across the country), and we have simultaneously created
their favored habitat—abundant shrubby plants and herbaceous perennials—
nearly everywhere.

It took a while, but throughout their range, deer populations are larger
than the environment can sustain without serious degradation—sometimes
several times larger than an area's carrying capacity. Deer support huge populations of black-legged ticks, which spread Lyme disease. Deer exacerbate the
invasive plant problem by eating native plants and not touching the invasives.
They have all but eliminated the understory of many of our forests and woodlots (except for said invasives), ending recruitment in these vital ecosystems
and seriously reducing populations of ground-nesting birds. Today, in far
too many places, when a mature tree dies, there is no seedling of its kind
to replace it naturally in the forest. Ultimately, we must bring deer populations back down below the carrying capacity—or else! Yes, that is a threat that
Mother Nature has clearly issued.

What to do? We could reintroduce deer predators such as black bears,
cougars, and wolves. It seems far-fetched to many folks, but this is actually
working in some areas. I recently visited Great Smoky Mountains National
Park and was surprised to see healthy and diverse understory growing
throughout the park. After inquiring how the rangers managed this, they
pointed out that they do not have a deer problem. They do have healthy
populations of white-tailed deer, but they also have healthy populations of
black bear, bobcats, and coyotes that keep deer populations in check. Black
bears are making a comeback in rural areas of North Jersey, Pennsylvania,
western Maryland, and West Virginia and may be starting to help control
deer populations. Where wolf populations have been eliminated or reduced,
coyotes have spread across the country, and in most suburban areas, they are
the only predators that can help control deer. Coyotes will take fawns for only
a few days after their birth, but this does help control deer populations. Unfortunately, hunting and killing coyotes is legal throughout the United States.

It is often suggested that we capture deer where they are too numerous
and take them to areas where they are uncommon. But this will not work, for

three reasons: First, deer are hard to catch. It takes professionals with rocket nets or tranquilizer guns, and this is a very expensive way to catch a few deer. Second, there are no places where deer are uncommon, and parks like the Smokies will not welcome ecological disaster by building their deer populations. Finally, if we could catch enough deer to meaningfully reduce the population in an area, it would take only one or two years for the population to rebound to its original size.

Many people suggest that a sterilization substance could be invented and darted into a deer. This sounds reasonable, but after 30 years of trying, no scientist has succeeded in creating a single-dose product that will effectively sterilize a deer. And good luck trying to get close enough to a deer to dart it twice. Many have tried. All have given up. Why not put something that will sterilize deer in their food? Another good idea that no one has mastered. Once eaten, the product is broken down in the deer's four-chambered stomach long before it can get into its bloodstream.

Why can't hunters control deer? Two reasons: There are fewer hunters every year, but the big reason is that hunters are not allowed to hunt within a set distance from housing developments. This is certainly reasonable, but it gives deer a safe place to hide. In most built areas, the only things a deer has to fear are cars. It is possible to hire sharpshooters for the sole purpose of reducing deer populations. These professionals work at night with night-vison glasses and laser target finders. They are good and efficient, but very expensive. Most communities don't consider springing for the cost until someone in the community dies in a car crash with a deer. Sharpshooters would work if they were employed everywhere all the time, but single efforts here and there are not enough. And public pressure *not* to control deer is enormous. It's called the Bambi effect, and we can blame Walt Disney for its prevalence, particularly among baby-boomers who all crowded into theaters as kids to watch cute little Bambi. I was one of those kids, so I do understand the emotional issues here. I doubt the public would protest as much as it does if Bambi looked like a warthog.

Bernd Blossey of Cornell University is convinced he has the answer to deer overpopulation, and I am convinced that he is right: market hunting. If hunters were allowed to hunt deer all year and sell the deer meat they harvested, there would be enough incentive to bring deer populations closer to ecologically sustainable levels. Right now, though, our hands are tied by existing regulations. Federal law clearly states that wild game species "cannot be sold." This law was enacted in response to market hunting in the late 1800s that eliminated the passenger pigeon, Carolina Parakeet, and Alaskan curlew, and all but wiped out bison, beavers, sea otters, egrets, ducks, geese, Sandhill Cranes, swans, prairie chickens, grouse, and other animals. With the end of market hunting in 1905, many of these beleaguered populations were able to recover beyond the threat of extinction, but today threats to biodiversity are coming not from too much exploitation but from too little. Blossey is not calling for wholesale market hunting of everything that moves, as was the case in the past, but for strategic market hunting of white-tailed deer wherever they exceed the carrying capacity.

Could you tell us how you handle deer on your own property?

While we wait for some concerted effort to bring deer numbers back down below the carrying capacity, there are three ways we can garden in their presence: use only plants that they don't like, constantly spray plants with a "deer-be-gone" smelly product, or fence (cage) them out. The problem with choosing plants that are unpalatable to deer is that most of those plants are also unpalatable to the insects we want to support. I don't like the deer repellant solution for two reasons: First, you have to be very vigilant and respray every plant after every rain. I am way too lazy for that. Also, when you make your plant so smelly it repels deer, you have probably changed its odor so much that the insects that need that plant won't recognize it (insects locate their host plants by odor, not sight). So I favor choosing the plants that are best at supporting insects (keystone plants) and then protecting them from deer.

My wife and I own 10 acres, way too much land to put a fence around, at least on our budget. So I bought wire fencing at a big box store (a roll of 5-by-150 feet steel mesh) and fashioned individual deer cages, which I place around every tree I planted. I made my cages liberally large, at least 4 feet in diameter, which gave the trees all the room they needed to grow. Then I staked each cage with a 5-foot piece of half-inch rebar. And that's it. When my trees grew past the height at which the deer could kill them, I removed the cages (graduation day) and used them on new plants. I have been moving the same cages around the yard now for 23 years and they are still good as new.

Wire cages staked with rebar successfully keep deer from browsing young oaks.

One word of caution: Even after a tree is safe from herbivory, deer can kill a young tree by buck rubbing. Every fall, bucks grow new antlers that are covered in velvet, the skin that actually produced the antlers. After the antlers are full grown, the velvet dies and must be scraped off. Bucks do this by rubbing their antlers up and down over the bark of trees with a diameter of 2 to 5 inches. This is abrasive to bark and can rub the bark right off the tree, killing the cambium and eventually the tree itself. It is easy to protect these trees by loosely wrapping the trunk with plastic fencing.

Another option that I have not tried but that I hear works well is to double-fence them out. Deer are willing to jump a single 8-foot fence, but they're very reluctant to jump two 3-foot tall, parallel fences that are separated by 3 feet. If this works, you can fence deer out of large areas with decorative split-rail fencing, a great option when aesthetics need to be considered.

Bucks will scrape the velvet off their antlers by rubbing young tree trunks.

To prevent this, wrap the trunk of vulnerable trees loosely with plastic deer fencing.

I am searching for full-sun, deer-resistant native shrubs. Can you help?

Deer love most native shrubs, but they avoid spicebush, buttonbush, silver-bell, fringe tree, and pawpaws, for starters. I understand your desire to use plants the deer won't eat, but if we use only deer-resistant plants, the food webs in our yards will collapse. The most productive plants are those the deer like the most, with oaks heading the list. One word of caution: A very hungry deer will eat anything. As I write, the deer at my house have eaten all of my young spicebushes. We have more deer than ever before, so I suspect they are all very hungry.

I live in an area with lots of Lyme disease. Several websites suggest keeping large lawns because they are not attractive to ticks. The sites also say I should get rid of brush piles because ticks love them. How can I include natives in my yard without getting Lyme disease?

Whenever I get this question, two platitudes immediately pop into my head: life is not risk-free, and life is a trade-off. It's true that lawns, whether mowed frequently or not, do not support a large tick population (or anything else), and pavement supports even fewer ticks. To create a world with no ticks, we could turn everything into lawn or just pave the world. The risk from Lyme disease would then drop to zero, but so would the probability that we will persist on this planet.

Let's think about what deer ticks need to complete their life cycle and then think about the easiest way to disrupt that life cycle. Deer ticks do not eat native plants, leaf litter, or brush piles. They stay in those areas because they need high humidity. Deer ticks feed on the blood of vertebrate animals, and two of their favorites (in fact, both are necessary to complete their life cycle) are white-footed mice and white-tailed deer. Although Lyme disease has always been around in very low frequencies, it reared its ugly head in the 1970s because white-tailed deer changed in abundance from rare sightings in the 1950s and '60s to several times over their carrying capacity. Too many deer are bad news for the environment for several reasons, and that seems like the best place to interrupt the Lyme disease cycle. We need to agree collectively as a society that an overabundance of deer is not OK, and it's not something we have to tolerate. I realize a single homeowner, however, can't bring deer populations down to ecologically safe levels, so what should you do in the meantime to minimize your exposure to deer ticks?

Deer ticks do not run after us when we go into our yards. They climb up on vegetation and quest—that is, they wait for us to walk by and then grab on when we do. So one easy solution is to reduce your lawn to wide-mowed paths, and then stay on those paths during periods of high tick infectivity (which is May and June in southeast Pennsylvania). For me, staying out of the woods is not an option, so I remain vigilant. I (with a little help from my wife) check myself for ticks after I've been playing outside. Deer ticks like bare patches of skin under waistbands, sock bands, or tight undies, or under your armpits, and they can be easy to find with close inspection. They also like to get between my toes. When I find an embedded tick, I pull it off (sometimes I need tweezers for those tiny nymphs) and put antibiotic ointment on the bite site. A Lyme researcher told me years ago that the ointment kills the Lyme *Borrelia* spirochete (bacterium) before it gets into the bloodstream if you apply it soon enough. I don't know if that is true, but I do know that I have not gotten Lyme disease when I follow that rule (and I have had it five times when I *didn't* follow the rule). This all may seem like more aggravation than it's worth, but the joys I get from interacting with nature far outweigh the nuisance of tick checks.

Interesting research spearheaded by Rick Ostfeld of the Cary Institute and Felicia Keesing from Bard College suggests another way to reduce Lyme disease risk. Create a landscape with lots of alternative tick hosts. When your landscape is so simple that it has only deer, mice, and you in it, disease transmission can be high. But when you add chipmunks, possums, foxes, raccoons, squirrels, and other animals, many infected ticks use those other mammals as hosts rather than you, and many of those mammals are dead-end hosts for Lyme disease.

A good example of this phenomenon can be seen in the South. There are too many deer and lots of deer ticks in the South—just like in the North—but the frequency of Lyme disease is very low. There are also lots of lizards of several species in the South. So what's the connection? You might think that the lizards are eating the ticks, but that's not it. Instead, the ticks use the lizards as hosts, but lizards do not support the *Borellia* bacterium life cycle, so lizards are dead-end hosts. They are effectively syphoning off infective ticks, lowering the risk in areas where there are lots of lizards. The point is that anyone who tries to reduce Lyme risk by removing nature from their yard will actually increase the risk of a Lyme infection.

A few controls can reduce the number of ticks. A proactive approach to controlling ticks that works pretty well is to deploy a product called Damminix tick tubes, cardboard tubes stuffed with cotton that has been laced with the insecticide permethrin. White-footed mice that host Lyme-infected ticks gather the cotton and build their nests with it. The cotton then kills the ticks that are on the mice without hurting the mice. There's also a naturally occurring fungus that's deadly to ticks and can be sprayed over large areas to reduce tick populations. If nothing else, these controls reduce Lyme infections in pets. Margaret Roach of *The New York Times* wrote a superb summary of tick research in 2021, which I highly recommend.

What native plants provide food for tick eaters?

Unfortunately, very few creatures eat deer ticks (they are simply too small), and those that do (a few small rodents) are not plant eaters; they are predators.

Can wildfires suppress tick populations? We never saw a tick at our cabin growing up, but in the last 15 years or so, they're everywhere. As kids, we played out in the fields tick-free, but now I can't sit at the picnic table without being mobbed.

Yes, fire suppresses tick populations. But what has caused the large tick populations we see now compared to when we were kids is an overabundance of deer, not the lack of fire.

I've read that acorn bumper crops can lead to upticks (no pun intended) in tick numbers. It seems in many ways that our efforts to rewild the urban landscape can have unwanted consequences. Is there a way to encourage natural predators that will keep this all in check?

Acorn masts do indirectly increase tick numbers by increasing the number of white-footed mice in the area of the mast. But let's not ignore the other aspect of acorn masting behavior in oaks. Following a mast year, oaks often produce very few acorns for the next several years. This causes the collapse of the mouse population and thus the subsequent collapse of the tick population. So averaged over the years, oaks do not increase the hosts of black-legged ticks. I would argue that, although rewilding the urban landscape does have some drawbacks, the benefits far outweigh them!

CHAPTER SEVEN

Conservation and Restoration

If you were able to give your ten-year-old self one piece of advice about conservation, what would it be?

I would tell my young self to think not only about preserving pristine habitat, but also about restoring Nature to the many places from which we have expelled her. When I was growing up in the 1950s and '60s, the impacts of our rapidly growing human footprint were obvious enough, but, like everybody else who was concerned about the loss of nature, I focused 100 percent on saving the bits of nature that had not yet fallen to the bulldozer. Not once did it occur to me that I could rebuild effective habitat right in my yard. I didn't know anything about native plants versus non-native plants, or about how many caterpillars were required to support breeding birds. But I did know that little ponds supported lots of very cool creatures, and I mourned the loss of the pond that had been bulldozed to make my neighbor's backyard. But why didn't I dig a new pond in my own backyard? My parents would not have minded; in fact, they probably would have helped me. Instead, I mindlessly mowed our yard each week, singing those great oldies that weren't oldies yet as I worked. This was a lost opportunity, and it would be four more decades before I realized that saving the nature that remained would not be enough to prevent ecosystem collapse. We need to restore nature where we live, work, shop, and farm, because we need functioning ecosystems everywhere.

When you're replacing invasive species with native trees, what happens to the understory? Do wildflowers and other native small plants revive? I read that once fields are plowed for crops (and probably for subdivisions as topsoil is carted away), the seed bank is lost. What has been your experience?

I believe the seed bank is far more resilient than many people think. Sure, much of the fantastic diversity of virgin Midwest prairies has been lost from the seed bank in the last 150 years, but much of it hung on in small "weed" patches that spread their seeds far and wide when given a chance. I have not yet inventoried the herbaceous plants that have survived on our property (I'm not a good enough botanist), but I know it amounts to hundreds of native species, and our property was farmed for 300 years before we bought it.

Your *Nature's Best Hope* plan is catching on in theory here on Whidbey Island in Puget Sound. But in practice, we face what appears to be a stumbling block. We would have no trouble giving up our lawns for native species. But we'd be hard-pressed to produce the butterflies to make the plan work. There are only about 10 species of regularly occurring butterflies here. If we see any more than a few of each of them annually, we are fortunate. To make matters worse, nearly all use grasses, broadleaf trees and shrubs, or species such as nettles as host plants. Given this mismatch, how should we define success?

Butterflies are a poor metric for the energy provided by caterpillars to food webs. Butterflies do not taste good and so contribute little to food webs. When you look at what food is brought to nestlings, butterflies represent less than 1 percent of their diet, whereas for most terrestrial bird species, moths dominate the nestling diet. For every species of butterfly, there are 19 species of moths. My guess is that you have good populations of moths such as the tufted thyatirid, the lunate zale, and the western eyed sphinx moth, all nocturnal species that are easy to miss if you don't search at night. It's not the number of species of moths that counts; it's the biomass of caterpillars generated by plants. For example, in many regions of the boreal forest, there are not very many species of moths. But single species such as the spruce budworm are so numerous that the boreal forest is a breeding destination for many neotropical migrant birds.

Do large restoration projects have to follow the sequence in which plant species normally enter a disturbed community?

No, they don't have to. It is very common to speed up secondary succession through tree and shrub plantings. We used to think that species not planted in the natural sequence of species that typically occurs during succession would not survive. But that has largely been disproven. Succession sequences are determined primarily by the mode of seed dispersal, not by the plant's ability to survive in an early successional habitat. Tulip poplars, for example, are often the first trees to colonize an abandoned corn field because their seeds are wind dispersed and reach the field first. But if a Blue Jay buries an acorn in that field and then forgets where it is, the acorn will germinate and the oak will do fine. Whether or not we restore ecosystems as nature would have often depends on our time scale and available resources: tree plantings cost money and require labor, but they do speed the restoration along, sometimes by decades. That doesn't mean you can't complete a successful restoration without a large labor force and lots of money. If we have a little money and lots of time, we can guide restorations simply by keeping out species we don't want. Addition by subtraction: remove invasive plants as they come in and let natural succession add the native species we hope to restore.

Would you please talk about micro forests and how close together to plant the trees? There was a piece on National Public Radio about micro forests, and they didn't say anything about how close to plant the trees, but they said it will take decades, instead of a century, for these forests to come to maturity.

Well, that is a bit overstated. The goal of a micro forest, a concept known as the Miyawaki method of afforestation, is to get plants and the animal diversity that uses those plants into a degraded space much more quickly than they would colonize naturally. It's not really about creating a forest. The micro part of a micro forest is its size: micro forests are planted in very small spaces that will be able to accommodate only a few mature trees in the end, and a few trees do not a forest make. But by heavily planting a small area with lots of young trees and shrubs spaced just a few feet apart, while simultaneously enriching the soil, you can achieve rapid growth in just a few years. Your micro forest will look great, and you will have attracted a lot of biodiversity where you previously had none. So far, so good. A few more years down the road, however, the plants, particularly the trees, will compete with one another for light, and the fastest growing individuals will win, shading out many of the others. Twenty years later, you will end up with one or two trees and little else. It's certainly not a mature forest, but remember, that wasn't really the goal to start with. If you start your micro forest with small plants, particularly those you have started from seed, the initial cost is minimal, and you will have supported much more biodiversity en route to your mature trees than you would if you had planted specimen trees from the start. Planting densely from the start will also limit weeding and the invasion of non-natives in the early years, two more important benefits of the Miyawaki method.

Do you support the restoration of the American chestnut?

I most certainly do! In huge swaths of the Appalachian Mountains, from Maine south to Georgia, the American chestnut was a dominant forest tree. Not only did it support hundreds of insect species, as well as the birds that needed those insects to reproduce, it also produced a reliable mast every year that bears, turkeys, deer, and many other critters depended on to make it through long winters. Moreover, chestnuts were an important food and wood source for Native Americans for many thousands of years, as well as for European colonists. All of the bounty provided by American chestnuts ended abruptly when we imported Asian chestnuts that were infected with the chestnut blight in the late 1800s. Unfortunately, at the same time, we began mail-order shipments of horticultural stock up and down the East Coast. So not only did we import a deadly disease for which American chestnuts had no resistance, but we also efficiently spread that disease throughout the range of the chestnut, resulting in billions of chestnut trees dying in just a few decades. Other than clear-cutting, the chestnut blight was probably the most destructive ecological challenge to deciduous forests since the Last Ice Age.

The American Chestnut Foundation was founded upon the premise that resistant genes from Asian chestnuts can be inserted into the American chestnut genome, either through backcrossing or genetic engineering, to produce American chestnut populations that can thrive in the presence of the chestnut blight. What's not to like about returning this productive tree to our beleaguered forests?

Is it possible to reintroduce an insect species to places from which it has disappeared?

Absolutely! And one of the best examples of this is the beautiful Atala butterfly. This species is host-specific on coontie (*Zamia integrifolia*), a cycad native to south Florida. Coontie roots contain a lot of starch—so much so that an entire industry was developed based on wild populations of coontie. "Starch gatherers" combed south Florida for coontie, very successfully I might add, for they "gathered" just about *all* the coontie, causing its local extirpation. Without its only food plant, the Atala also disappeared and was presumed extinct from Florida in the mid-1960s. In 1979, however, Roger Hammer, a park manager for Miami-Dade Parks, discovered a remnant population of Atalas on a small surviving population of coontie on Virginia Key. This was at a time when coontie had become a favored landscape plant and was no longer being harvested for starch in the wild. He reared and distributed Atalas in a number of protected sites, where it thrived and expanded on its own. Today Atala butterflies are common in most of south Florida where coontie has been welcomed back into the landscape. Insect reintroductions are among the easiest and most successful, as long as we provide what the target species needs.

The stunning Atala butterfly was saved from extinction in Florida by returning its cycad host plant, coontie, to garden landscapes.

Did you choose plant species you wanted for your yard based on what insects they would attract, or did your plant choices reflect what made sense for the natural environment and history of the area?

Long ago I learned that plants are much more than decorations, so I adjusted my plant choices (when I did make plant choices) to those that offered the most benefits to my local ecosystem. In other words, I selected plants that hosted the most caterpillar species, supported the most specialist pollinators, removed the most carbon dioxide from the atmosphere, and protected our watershed the best (that is, had the largest root systems). This was not a big sacrifice, because the life those plants have brought to our property has given me far more pleasure than just another inert pretty plant. But if I were to inventory the plants that are now present in our yard, I would discover that I actually planted just a small percentage of them (20 percent). The rest have colonized by natural means.

How long will it take a new native plant garden to reach a balance between herbivores and predators so that insects won't defoliate my plants?

It doesn't take long at all, assuming you have provided what each trophic level (plant eaters, predators, parasitoids, and diseases) requires to make a living. Insects colonize plants quickly, assuming there are source populations nearby. And it doesn't take long for the birds and other natural enemies to track down those insects. I think a better question would be, how many native plants do I need to have in my landscape to attract enough insect herbivores to sustain the natural enemies that control them? This is definitely a case where bigger is better. The larger and more diverse your planting is, the more powerful the chemical signal produced by the plants in your garden, so more insects seeking host plants are attracted to your garden. This assumes, of course, that you, your neighbors, or your township is not employing a mosquito fogging company. A single fogging event will kill all or most of the insects in your landscape, including the natural enemies of herbivore pests.

What single thing could home gardeners do to start sharing their yards with the natural world?

Plant an oak tree! Oaks support more biodiversity, sequester more carbon, and manage the watershed better than any other tree in most parts of North

America. They also support pollinators, even though oaks are wind pollinated. Bees regularly go to oak catkins and gather pollen, which they then
feed to their young. Start with the smallest tree you can find (acorns are free!).
You will end up with a faster growing, healthier tree this way.

Is it better to get more regional plants (300–400 mile radius) at a cheaper price or buy from local native plant nurseries (75–125 mile radius)?

The answer depends on why you are asking the question. If you are wondering whether a plant purchased close to home will do a better job of supporting
the food web than a plant purchased a few hundred miles away, the answer is
probably no. Food webs are more forgiving about the provenance of a plant
(the location to which that plant is adapted) than most plants are. For example, an insect that specializes on American beech does not care if the leaves it
is eating come from a beech genetically adapted to northern Florida or from
a beech adapted to southern Canada. The leaf chemistry, which is what determines host use in insects, will be so similar between the two beech options
that a beech specialist will not be able to tell the difference. However, there is
some chance that a plant genetically adapted to a latitude 400 miles south of
you may struggle during a tough winter in your yard. Supporting your local
nursery is always a good idea in terms of supporting your local economy, but
you should check to be sure where the nursery's seed sources are located.
Many nurseries source their plants from farther away than you might think.

Is it better to have fewer plants of lots of species or more plants of fewer species?

The answer depends on your restoration goals, the species you are trying to
restore, and the size of your property. In general, diversity is a good thing.
American ecologist Robert MacArthur rightly predicted way back in 1955 that
the stability and productivity of an ecosystem would be a function of the number of interacting species in that ecosystem: the greater the number of species,
the greater the stability and productivity. But we have learned a lot since 1955
that refines his general prediction. For example, we now know that not all
species are equally productive, so we can increase ecosystem productivity at
least somewhat by favoring the most productive species. That would reduce
overall diversity a bit. All restorations should be able to sustain themselves,

which means the plants in that restoration need to be able to reproduce. If you are including dioecious species (male and female flowers on different plants) such as boxelder, spicebush, and various hollies, you need to have both male and female plants in the population or they won't be able to reproduce. So again, rather than planting one individual of each species, you would need fewer species with more individuals of each. Wind-pollinated species such as grasses, oaks, maples, and pines produce much more seed when they are in large populations because pollination success depends on the amount of pollen that is released. Finally, if your restoration is targeting a specific species, such as the monarch butterfly, for example, you would want to plant milkweed patches containing a number of individuals as well as fall bloomers such as asters and goldenrods to provide the nectar for migrating monarchs. On small properties, those requirements might easily use up all of your available space, compromising your ability to add other species.

I've added a number of new native species to my garden this spring, including keystone plants like pussy willow and goldenrod. I live in the suburbs with little in the way of natural habitat nearby. Will it take new species of insects a year or two (or longer) to find my garden?

You would be surprised how fast at least some insects find their host plants. It will depend on how far they have to travel to locate your new plantings, but I regularly find moths on my property whose host plants are at least a mile away. Years ago I planted pawpaws to attract zebra swallowtails. The nearest population of zebra swallowtails that I knew of was 27 miles away. They did find my pawpaws, but it took them nine years to do it. In general, larger insects travel farther and faster than tiny ones. When I was a student at Rutgers University, we would find sphinx moths from South America in light traps several times each summer. So, short answer: some species will find your new plantings the same year you plant them, others will take a bit longer. But they *will* find them.

I am not the only one measuring rapid increases in insect populations following home restoration projects. A recent study in Australia by Luis Mata and others looking at the ecological outcomes of urban green spaces found that even small greening actions resulted in large increases in insect abundance, diversity, and plant/insect interactions in only three years. As much as I have grown to hate the cliché, it is true: if you build it they will come!

How should I restore my understory after extermination of extensive buckthorn and garlic mustard so that deer don't destroy the new plants?

The ugly truth is, until we get our deer populations back down below the carrying capacity, you will need to protect your plants against deer. First, I wouldn't worry about garlic mustard. Bernd Blossey's work at Cornell suggests that, when left alone, populations of garlic mustard decline over time. Every time you pull one from the soil, you reset the clock by disturbing the seed bank. But buckthorn invasions are almost always extensive and, when removed, leave a blank slate, ripe for reinvasion. You can actively replant canopy trees, understory trees, and shrubs, but most will need to be caged with 5-foot-high wire mesh (mesh size about 2-by-3 inches) fashioned into cages about 4 feet in diameter and staked with rebar. Once built, those cages last forever and can be moved from plant to plant as needed. Being thrifty by nature, I don't buy plants for my restoration projects; I just protect those that come in on their own. Jays and squirrels alone will plant lots of oaks, hazelnuts, walnuts, hickories, and beeches for you, and songbirds will poop out many more types of seeds, including those all-important viburnums. Your big challenge will be removing new invasives as they inevitably come in. A sweep through the site with a mattock once or twice during spring and summer will be enough to keep your site free of young buckthorn, multiflora rose, oriental bittersweet, porcelain berry, and other invasive plants.

In my remnant bottomland forest, does it help wildlife more if I remove the maples, sweetgum, and tulip poplars and replace them with cottonwoods, black willows, and sycamores? There are trees in a ditch at the front of my property I can transplant before they get mowed again by the county.

When logistically possible, it is always better to push the species composition of a forest patch more toward keystone species (those that contribute the most energy to local food webs). It is common, particularly in the eastern coastal plane, to have an overabundance of maples, tulip poplars, and sweetgum, so replacing at least some with black willows and cottonwoods, with an occasional sycamore thrown in, sounds like a great idea.

What is the best method to fill in the spaces where ash trees have died— plant seedlings or seeds (nuts)?

This question is very similar to an earlier one about restoring spaces after invasive plants are removed. The big difference is that we need to encourage ash seedlings after the emerald ash borer has moved through and killed mature ashes. The borer cannot develop in very young ashes (the trunks and branches are too skinny), so areas where it has killed mature trees are often well-endowed with ash seedlings. These should be encouraged as much as possible. Why bother if the borer will just kill them when they get larger? Because great strides have been made with biological control of the ash borer. At least three species of parasitic wasps imported from Asia have become established and are starting to reduce ash borer populations in a number of sites—a rare bit of good news in the world of invasive species.

If you worry that biocontrol agents will often switch hosts and start killing native species, worry not! These days, nothing can be imported to the United States without years (often a decade or more) of host-specificity testing in USDA quarantine facilities. These parasitoids are specialists on ash borer, and only ash borer. Once established, they will do a regulatory dance with the borers, building in numbers when the borer populations increase and shrinking when the borers decline. The emerald ash borer will never be eliminated from North America, but these natural enemies should keep its populations low enough that ashes will be able to tolerate low-level attacks.

I recently heard that a California court ruled that bumble bees are fish. Please explain.

It's true! In 2022, the justices at California's Third District Court of Appeals ruled that bumble bees are fish and are therefore vertebrates that have the legal status and protection of fish and other endangered vertebrates in the state. This, of course, was merely legal gymnastics required to extend protection under the California Endangered Species Act to bumble bees and presumably other pollinators. Two things stand out about this decision: first, the fact that it was necessary at all and, second, that the court ruled in favor of protecting bumble bees. The decision to extend protection to bumble bees and other invertebrates reversed a lower court ruling that insects do not deserve protection from the Endangered Species Act. And that, in a nutshell, explains why planet Earth is experiencing global insect decline and has already lost more than 45 percent of the insects that run the ecosystems on which we humans depend.

Despite E. O. Wilson's well-known 1987 declaration that insects are "the little things that run the world" and the fact that humans would survive only a matter of months in a world without insects, we have waged a successful war against our six-legged benefactors ever since we invented agriculture. The fact that a higher court has recognized the ecological value of insects and other invertebrates, even if they have to declare that insects are fish to do it, is real progress.

President Biden wants to protect 30 percent of the country. Is that enough?

Actually, President Biden is supporting the United Nations' resolution to protect nature on 30 percent of the planet by 2030 (30×30), and 50 percent by 2050 (50×50). This goal comes from E. O. Wilson's 2016 book, *Half Earth: Our Planet's Fight for Life*, in which he explains that to sustain life anywhere on Earth, we need to have functioning ecosystems (nature) on at least half of the planet. The real question is, how can we save nature on half of planet Earth when half of terrestrial earth is in some form of agriculture, and we have eight billion people, and all our infrastructure is in the other half? Short of getting rid of people, the only way I see to meet Wilson's goal—that is, to ensure human survival in the future—is to find ways for humans and nature to coexist, in the same place, at the same time. Only 3.6 percent of all land in the United States is protected in the national park system, and only 12 percent has any federal protection at all, so to meet the 30×30 and then 50×50 objectives, we are going to have to get serious about conservation on private property—all of that land outside of parks and preserves, including our yards, corporate landscapes, roadsides, powerline cuts, golf courses, and even much of our agricultural acreage.

What can agriculture do to increase insect populations?

A lot! The United States has 410 million acres of cropland, 406 million acres of woodlots managed by private citizens (not logging companies), and 770 million acres of rangeland. That equals more than 1.5 billion acres devoted to some form of agriculture. There are more than 2.26 billion acres of land in the country, which means about 66 percent of it is in some form of agriculture. When viewed in this way, the need to conserve insect populations on agricultural lands becomes painfully obvious.

Let's first consider cropland, the area most people think of when they think "agriculture." There are four ways to increase insect populations in cropland: roadside management, prairie strips, hedgerows, and the elimination of needless insecticides. The first and perhaps most important measure is to restore insect habitat along roadsides within croplands. Literally hundreds of thousands of miles of roads crisscross our croplands. On either side of each of those roads are two strips of land, or verges, each typically at least 30 feet wide, between the road and the actual crop. In the past, all of this acreage was dominated by native perennial and annual plants, such as milkweeds, goldenrods, asters, evening primrose, perennial sunflowers, and many remnant prairie species—you know, the plants most folks call weeds! And these were the plants that supported robust populations of native bees, grasshoppers, caterpillars, and our beleaguered monarch butterflies. Cropland coexisted with these roadside insects, none of which were actual pests in the crops themselves, for well more than a century in the Midwest and for more than three centuries in the East. The largest monarch populations ever recorded occurred in the mid-1970s and were produced largely in cropland roadside verges.

Then roundup-ready corn and soybeans were invented. That enabled growers to spray their crops to kill weeds without hurting the crop itself. But it also enabled them to spray the land right up to the roadside, eliminating the plants that supported most of the life in farmlands. What happened then? They planted all of that acreage in turf grass! Roadside turf grass quickly became the new agricultural status symbol. All growers worth their salt could advertise their status by planting lawn on verges throughout their holdings. No matter that all of that lawn then needed to be mowed regularly; it looks nice and neat, and it now dominates cropland roadsides from here to eternity.

Across much of the United States, cropland verges that were once home to monarchs and pollinators have been planted in turf grass, resulting in significant declines in populations of those important insects.

Immediately, monarch populations started to tumble, and there is little doubt that all of the other butterflies, moths, and native bees that depended on roadside vegetation also declined, even though no one has monitored their populations. Their habitat was eliminated.

In this case, the fix is obvious. Put the plants back! Removing roadside vegetation has not benefited crop yield, and removing the need to mow roadsides will actually save growers time and money. We need to flip the status symbol, so that growers who maintain lawn verges will become known as the uninformed destroyers of monarchs and pollinators. And this, happily, is actually starting to happen! Iowa, for example, is leading the way in the use of natives in roadside planting. The Iowa Department of Transportation manages more than 55,000 acres of roadsides planted with coneflower, milkweed, penstemon, blazing star, and dozens of other species of trees, grasses, shrubs, and wildflowers.

There is a second way we can restore insect populations to agricultural lands. Hedgerows! In the last 20 years or so, there has been pressure to remove hedgerows from agricultural land to make harvesting with giant machines easier. In my view, this movement got out of hand and hedge-rows were removed even when they weren't in the way, particularly along roadsides. Quite simply, restoring hedgerows and maintaining them free of invasive plants would be an excellent way to boost insect (and bird) diversity in agriculture.

A new and powerful way to add insect habitat to agriculture is through the strategic use of prairie strips, which are just what they sound like: strips of diverse prairie plantings right through corn, soybean, and wheat fields. They are oriented perpendicular to the flow of water off of a field so that they can intercept topsoil and excess nutrients *before* these things reach nearby streams and rivers. Research at Iowa State University spearheaded by Lisa Schulte Moore and Randy Kolka has demonstrated remarkable benefits associated with prairie strips. Increasing forage for pollinators and other flower visi-tors is an obvious benefit that would dramatically boost insect populations wherever they are employed. Less obvious is that prairie strips reduce storm-water runoff by 44 percent, topsoil loss by 95 percent, phosphorus runoff by

Prairie strips help pollinators, intercept topsoil, and reduce watershed pollution.

90 percent, and nitrogen runoff by 84 percent. Compared with catchments containing only crops, prairie strips increased insect diversity more than 2.5 times, increased pollinator abundance 3.5 times, and more than doubled native bird diversity and abundance. Fortunately, the USDA Conservation Reserve Program (CRP) recognizes these incredible benefits and will help offset the costs of prairie strips for growers who install them. This is a win-win for growers, soil conservation, water pollution, and biodiversity.

Finally, a commonsense way to help insects in agricultural lands is to stop killing them needlessly outside of the crop. I'm talking about neonicotinoid seed coatings, that pink stuff covering nearly every corn, wheat, and soybean seed sold that is designed to kill whatever might munch on these crops. Through some incredible bureaucratic reasoning, seed coatings aren't even labeled as insecticides and therefore are not subject to regulation, even though they are applied to crops on more than 150 million acres each year, according to the Environmental Protection Agency. Regardless of what we call them, neonic seed coatings are more than 7000 times more toxic to insects than DDT, and we use tons of this product in the United States each year (between 1994 and 2009, the United States deployed 14.6 million pounds of neonicotinoids). There are other features of seed coatings that are hard to fathom. First, only 2 to 5 percent of the coating is actually absorbed by the plant when it germinates. That means 95 to 98 percent either washes off into our watersheds, where it is extremely persistent, or blows away in dust who knows how far from the field in which it was applied. Second, it is used preventively—that is, it's used whether or not a grower has an insect problem. Finally, and most gobsmacking of all, studies have shown no increase in yield when neonic seed coatings are applied. Montana author/researcher Judy Hoy has also found evidence that neonics alone or in combination with exposure to glyphosate cause birth defects in mammals, including white-tailed deer and humans. To summarize, each year growers apply millions of pounds of a very potent insecticide that does not benefit growers through increased yields, but it does pollute terrestrial and aquatic habitats in ways we are just starting to understand! (There are not enough exclamation points to apply to that sentence.) Perhaps this is why most forms of neonicotinoid insecticides are banned in Europe!

How can we urge local, state, and federal governments to view trees as infrastructure?

I would expand this question to read, "How can we urge local, state, and federal governments to view trees as infrastructure essential to the common good?" The plainest answer is to help those of us impacted by local, state, and federal governments (that is, everyone) to understand just how essential trees and other plants are to their well-being. In a democracy, it is "the people" who determine who is making policy at the local, state, and federal level, and we do this every time we go (or don't go) to the voting booth. If the quality of local ecosystems, the biodiversity crises, or the environment in general were important issues for most people, it would at least show up among the top 10 concerns of the public. That has never happened, so we all have more work to do.

Any thoughts on how we can get nurseries to be more ethical and play a bigger role in environmental stewardship? It seems as if the big ones only want to sell non-native ornamentals and few provide education about the importance of native plants.

Nursery owners are in the business of selling plants; many of them don't care too much about which plants they sell, as long as there is a market for those plants. In the past, most nurseries carried few or no native plants because very few people bought them. That is now changing; the demand for natives exceeds the supply in many areas of the country. That's a business opportunity. If your local nursery is not up to speed on this cultural shift, you can help them out. Go there and ask for whatever native you are seeking. If they don't carry it, ask if they can get it for you. If they say no, then leave and don't buy something else. That is a lost sale and the owner will take note.

Monarchs

Why do people care so much about monarchs?

The monarch butterfly has long fascinated us. It is large and beautiful, and historically it was one of our most common butterflies. The monarch's most interesting feature, though, is its migration. East of the Rocky Mountains, millions of monarchs fly south each fall from Canada to the oyamel fir forest in the mountains in Central Mexico; in California, a similar north-to-south flight terminates in coastal pines near Santa Cruz and San Diego. But like so many other creatures, the monarch is in trouble. Destruction of its breeding habitats in North America and its overwintering sites in Mexico have caused such a precipitous decline in monarch numbers that the International Union for Conservation of Nature recently listed the migrating populations of this species as endangered. The good news is that groups across the country, coordinated by the Xerces Society for Invertebrate Conservation, Monarch Watch, Save Our Monarchs, and others, are planting milkweeds and fall forage in yards and waystations in a national effort to bring the monarch back from the brink.

I have heard that tropical milkweed may be bad for migration patterns. Should I stop planting it?

Although this was hypothesized, little evidence suggests that tropical milkweed impacts migration behavior in eastern monarchs. But this plant can pose a threat to monarchs in the form of a disease, at least when it is grown far enough south that it does not die back in winter. *Ophryocystis elektroscirrha* (OE for short) is a protozoan parasite specific to monarch and queen butterflies that causes wing deformation and ultimately death. Spores of OE accumulate on tropical milkweed tissues that do not die back in winter to the point at which 90 percent of the monarchs that visit a tropical milkweed plant get infected. So it is definitely a bad idea in Florida and along the Gulf Coast to plant tropical milkweed. Monarchs seem to prefer tropical milkweed not because it tastes better than temperate zone milkweed species, but because they prefer tender tissues of milkweeds, and tropical milkweed is always tender. Leaves of native milkweed species toughen up as the season progresses. A good way to give monarchs tender milkweeds in August is to cut some of your swamp and common milkweed plants back to the ground in mid-June. They will regrow and be nice and tender later in summer when most monarch reproduction occurs in the north. So is tropical milkweed a bad idea in Virginia? Let's just say it's not as bad as it is in the Deep South. But I always want to favor the native species when I can.

Tropical milkweed, *Asclepias curassavica*, is pretty, but plants growing in the South can infect monarchs with a deadly protozoan disease.

What is the best way to plant and propagate milkweed?

The most difficult way to start new milkweed plants is by seed. Seeds are easy to collect in autumn, but many are not viable; they need a cold stratification to germinate, and when they do germinate, the tiny plants must be coddled until they get big enough to transplant. I favor starting new plants from root stocks. Milkweeds are stoloniferous, meaning they send out lateral roots from which new milkweed ramets (stalks) grow. A milkweed patch is often one single plant with many ramets. You can quickly start new milkweed plants from sections of those lateral roots that are no bigger than an inch or two in length. Dig up a milkweed plant and divide its roots into 2-inch sections. Then plant each root sideways about an inch under the soil. In a few weeks, a new ramet will emerge and will grow far faster than a tiny seedling.

One of the plants I have is common milkweed. I had four plants last year, spread individually. There are larger clusters growing this year, which is really exciting. But they're very close to one another, and I'm wondering if their roots will crowd the others. I'm not sure if I should leave the plants as is or if I should move them to spread them out in my plot.

What you have is really one plant, rather than several crowded plants. Each of your stalks is connected to others by roots. I would leave them all alone.

I grow pink and white swamp milkweed to improve habitat conditions for monarch butterflies. I've seen names like 'Cinderella' for pink flowers and 'Ice Ballet' and 'Milkmaid' for white and wonder if this is just marketing or indeed different plants than swamp milkweed?

Those are different cultivars of the same species. A cultivar is a genetic variant of the straight species. In other words, 'Cinderella' is a genotype of *Asclepias incarnata* that expresses only pink flowers. It is just a slice of the wild genotype. If you're trying to help monarchs, it would be better to stick with straight species and avoid cultivars. Cultivars are selected to make people, not monarchs, happy.

Why aren't monarchs good pollinators?

Bees are good pollinators in part because they are designed for it. The branched hairs on a bee's body and legs pick up a positive charge while the bee is flying. Pollen, in contrast, has a slight negative charge, so when a bee lands on a flower, the pollen clings to its hairs. Monarchs and other butterflies are not built for pollinating. They do not have long, branched hairs on their bodies and their legs usually have no hairs. I have heard many people say that we need to save the monarch because it is a pollinator. In other words, because monarchs are helping us by pollinating plants, they are worth saving. The reality is, however, that although the monarch is a flower visitor, it is not very good at moving pollen from the male parts of flowers to the female parts of flowers. Does that disqualify it from conservation efforts? I certainly hope not. It deserves its place among the millions of other pieces of the great biodiversity jigsaw puzzle we call life, and without it, there would be an ugly hole in that puzzle. Though many may think so, I have never believed that our fellow earthlings exist solely to meet human needs.

I save every monarch caterpillar I can and kill all the assassin bugs I find because they prey on them. I have learned that assassin bugs are eating lanternfly nymphs and I am wondering, should I still be killing the predators? What do you think?

Don't kill the predators! They are a necessary part of every ecosystem because they keep herbivores in check. The best way to help monarchs is to expand your milkweed plantings. With lots of milkweed, it is easy to produce many more monarchs than assassin bugs can eat. This is one case in which we need to let nature take its course.

The monarchs on my milkweed keep disappearing. Just when they get to be big caterpillars, they disappear. I assume birds are eating them, but I thought monarch caterpillars tasted bad, so I'm confused. What is stealing my monarchs?

No, birds aren't eating your monarchs. Your monarch caterpillars are disappearing from your milkweed because they have finished growing and have crawled off to form their chrysalises some place away from their host plant. Monarchs, like most other Lepidoptera, distance themselves from their host plant before transforming to a chrysalis or pupa because they inherently "know" that their insect enemies, species such as predatory stink bugs, assassin bugs, tachinid flies, and many species of parasitoid wasps, search for their victims by first finding their host plant. If your monarchs are not on milkweed during the eight to twelve days of their chrysalis stage, they are more difficult to find.

Caterpillars can travel a good distance before pupating. I once found a pipevine swallowtail chrysalis hanging from a picture frame in our living room. The pipevine that had produced that caterpillar was 75 feet away. That caterpillar had crawled off the pipevine, across our backyard, up our back deck stairs, through our open door, across the dining room, and up the wall to its final resting place, the bottom of a picture frame.

Is rearing monarchs a good or bad idea?

Emma Pelton of the Xerces Society has written a thoughtful discussion about this controversial subject, but, in a nutshell, whether rearing monarchs does more harm than good depends on your intentions and the number of monarchs you rear. There are three reasons people rear monarchs: for fun, for education, and for conservation. There is little harm in rearing a few monarchs each year for fun. Rearing moths and butterflies is enormously entertaining, at least to nature nerds like me. I do it every year and find it to be a wonderful way to reduce stress (and avoid email). Giving the caterpillars new leaves and removing the old ones, cleaning the frass (caterpillar poop) from the containers, watching the little guys molt and grow, often changing color as they go, is surprisingly therapeutic, and I firmly believe that this daily exercise can cure anyone of depression.

There is general consensus that rearing monarchs for educational purposes is also worthwhile. So much can be learned by directly observing the life cycle of a monarch. Host plant specialization, one of the most important features of plant/insect interactions, is abundantly clear when you're rearing monarchs, since monarch caterpillars will thrive only on milkweed leaves. Complete metamorphosis, the exoskeleton and jointed appendages that characterize all arthropods, and the three pairs of legs and three body regions that characterize only insects, are on display for both young and old who have never thought much about insects, and it never fails to instruct and fascinate.

More and more, however, people with good intentions are trying to rear monarchs by the hundreds or even thousands to help reverse their declining numbers. This is where several problems arise. The motivation of people who rear large numbers of monarchs is easy to understand: monarch eggs and larvae suffer very high mortality in the wild. My own student, Brian Cutting, once showed that 90 percent of monarch eggs are carried off by ants before they even hatch. If kind-hearted people collect eggs, bring them indoors, and rear them through to adults, mortality is greatly reduced and many monarchs that would otherwise have fallen prey to ants and other predators can be released back into the wild, boosting monarch populations wherever this is done. That's the idea, anyway. There are two main problems with this otherwise logical approach to monarch conservation, however. First, when monarchs are reared in groups, diseases easily spread among the caterpillars. The protozoan *Ophryocystis elektroscirrha*, in particular, is easily spread as spores on emerging adults that may not themselves exhibit symptoms. Even less apparent, though, is that by bringing monarchs indoors and protecting them from the rigors of the environment, we deprive them of natural selection for the strongest, most fit individuals. Caterpillars that would have otherwise succumbed to any one of several environmental stressors are "saved," but so are their inferior genes. When released into the natural population, these less fit individuals dilute the genetic composition of those that have survived on their own. This may be why no one has been able to measure a bump in monarch populations from mass rearing attempts.

Bottom line is this: raising a few monarchs for fun or education is fine, but rearing large numbers of them should be avoided. The best way to help monarchs is to provide the resources they need. Monarchs have declined precipitously because we have replaced the milkweeds they need for development and the native flowering plants they need to fuel their amazing migration (asters, goldenrods, and so on) with lawns, primarily turf grass planted along the edges of agricultural fields wherever corn and soybeans are grown. There is no way the monarch will recover to reasonable population levels unless we add the plants they need back into our landscapes.

How many milkweeds do I need to support monarch breeding?

At a minimum, you need enough to support the complete growth of a single monarch caterpillar. On average, that amounts to two complete milkweed ramets (depending on their size). I once received an email from a woman who said she planted *a* milkweed plant for the monarch, but worms got on it so she squished them! The "worms," of course, were monarch larvae! The thing is, if you have just one or two milkweed ramets, a female monarch is likely to lay more than one egg on them because she can find no other milkweeds around. Those eggs will hatch into hungry caterpillars that very likely will eat all the milkweed leaves before completing their development. The milkweed will regrow, but the caterpillars will starve to death. So how much milkweed do you need? You need enough so that that a female can lay all the eggs she wants without her larvae running out of food—that is, a good-sized patch of milkweed. Fortunately, milkweed patches grow larger each year, so all you have to do is encourage them.

Whenever possible, plant a patch of milkweed rather than a single plant.

We have had a summer cottage on the eastern end of Lake Ontario for 43 years. Monarch butterflies used to cross the lake in quantity in summer, but now only a few cross. Hoping to encourage them, we planted butterfly bushes, which butterflies seem to like, as do the hummingbirds. Are butterfly bushes native to the United States?

Well, monarchs need different plants at different times of the year. During early and midsummer, they need milkweeds for reproduction and nectar sources for energy. In late summer and September, they have started their migration, and that's when they need fall nectar sources. Monarchs crossing the lake are migrating and no longer need milkweeds, but they really need fall bloomers such as goldenrod and fall asters. Butterfly bush is a good source of nectar in late July and early August, but it is not native and is becoming invasive. It does supply nectar, but not at the times when monarchs need it most.

What are those orange-and-black bugs on my milkweed? Should I kill them?

I wish we could learn to assume an insect is innocent until proven guilty instead of the other way around. There are two kinds of orange-and-black "bugs" on milkweed plants. Neither one is "bad." The first, the large milkweed bug, truly is a bug, a heteropteran for which the term *bug* is correctly applied. Like monarchs, they migrate to track the maturation of their primary food—in this case, milkweed seeds. Unlike monarchs, however, they do not fly to Mexico to spend winter; instead, they fly south to the Gulf States, where milkweed seed pods are available all winter long. In spring, they move north slowly as milkweeds start to form new pods from south to north. In the mid-Atlantic States and north, milkweed pods don't begin to mature until August, and that is when you first see large milkweed bugs in those areas. You should admire their many specialized adaptations for making a living on a toxic plant; you should *NOT* kill them.

The second orange-and-black "bug" that is common on milkweed plants, the milkweed tussock moth, is not a bug at all, but a lepidopteran that feeds gregariously as a caterpillar. Like the milkweed bug, it sports aposematic warning coloration, but in the caterpillar stage; as an adult, is a rather bland moth. It is difficult to convince people to leave these creatures alone because when they are present, there are usually many of them and they are voracious, often stripping an entire milkweed ramet of its leaves. Do milkweed tussock moths deprive monarch caterpillars of the tender milkweed leaves they need? Not usually, because milkweed tussock moth caterpillars seem to prefer older plants with tougher leaves. This may explain why they feed in groups rather than as individuals: group feeding seems to overcome the obstacles posed by tough leaves. Again, it is far better to enjoy the diversity of life associated with your milkweed patch than to contemplate killing it.

I hear about efforts to plant roadsides with pollinator plants to help bees and the monarch, but I worry that passing cars will kill these creatures. Is that a legitimate concern?

Unfortunately, that *is* a legitimate concern if the roads bear high-speed traffic. I did a study a few years ago with my student William Keilson, in which we

Milkweed bugs (adults at left, juveniles at right) are true bugs that specialize on milkweed seeds.

As a caterpillar, the milkweed tussock moth is quite beautiful, but the adult is rather bland.

compared insect mortality on roads bordered by grass versus roads bordered by meadows or woodlots. We found that insects *are* being killed by passing cars, especially bumble bees and dragonflies—two groups we don't want to lose. Somewhat surprisingly, though, insect mortality was not higher along meadows than along mowed turf. Apparently, insects have no reason to stay in mowed verges, so they fly through them, often right into traffic. When meadows border roads, however, insects tend to stay in the meadow plantings once there, but there are risks associated with reaching roadside meadows without getting squished by oncoming cars. Roads bordered by woods had the lowest levels of insect mortality. One very important takeaway from this study was that highways with well-planted median strips had the highest levels of insect mortality. There is no way a bee or butterfly can reach the flowering plants in a median without flying across traffic.

These results should not be interpreted in ways that discourage roadside restorations, but they can help guide how such restorations should be built. Many questions about roadside plantings should be investigated as soon as possible. All of our study sites had relatively high volume, high-speed traffic. How might these results differ in areas with slower traffic and less traffic volume, such as the thousands of miles of roadways crisscrossing Midwestern agricultural areas? Would insect mortality decrease if roadside plantings were one or two mower widths from the edges of the road? It was clear to us that regardless of the type of planting along roads, cars do kill insects. When properly designed and placed, can roadside plantings generate more pollinators and other insects than they kill? This would be a difficult experiment to design and execute, but it needs to be done so that we can take conservation advantage of the acreage bordering the four million miles of roads in this country in ways that do not create ecological traps.

Homegrown National Park

I would like to know more about Homegrown National Park than the info posted on HomegrownNationalPark.org. What is its mission and how did it come about?

In 2021, Michelle Alfandari and I started Homegrown National Park (HNP) as a non-profit to help get the message out that the future of conservation will be on private property and that we, all of us, are responsible for good Earth stewardship to go viral. The impetus behind creating HNP was to enlist a massive army of many millions of people—the public—in the fight to save biodiversity and help mitigate climate change. When I first learned that we had some 40 million acres of lawn in the United States, which is an area larger than all of New England, I saw an opportunity rather than an ecological disaster. Well, I *did* see the ecological disaster of transforming so much land into an ecological dead zone, but the opportunity to change an enormous restoration imperative into something manageable for everyday individual citizens was obvious. If everybody with lawn strove to reduce their lawn area by adding productive native plants to their landscapes, the job of turning millions of acres into viable habitat for birds, bees, and more would be divvied up among the multitudes of landowners across the country.

There are countless examples of success when people do replace some lawn with oaks, cherries, willows, birches, cottonwoods, goldenrods, asters, sunflowers, and other native plants. The only problem was that those millions of newly minted conservationists didn't know they were newly minted conservationists. They didn't know there was a biodiversity crisis every bit as important as the climate crises. They didn't know their landscapes were contributing to that crisis and that they, as individuals, had the power to reverse biodiversity declines. And they didn't know that along with "owning" a piece of the earth comes the responsibility of being a good steward of that piece. The fact is, every person on the planet has a responsibility for taking care of the ecosystems that support them, because every person on the planet depends entirely on those ecosystems for their well-being.

Michelle approached me to ascertain if I was interested in scaling my message. She was unaware of the biodiversity crisis until she heard me speak—in fact, she didn't know who I was before attending one of my talks at the request of a friend. She was not a gardener or a conservationist. She was a businesswoman from Manhattan! But she felt that if *she* could be motivated by my message, so could millions of others who were unaware of the issue and solution.

Thus, we created HNP, a small non-profit whose only mission is to get those important messages to the millions who need to hear them. Membership is free and therefore does not draw people away from existing conservation organizations such as Audubon, National Wildlife Federation, Sierra Club, Wild Ones, and others. We ask that people register "on the map" their property's location and the area of land they are or are planning to restore. Their tiny speck of land then "lights up" with a little firefly icon on the HNP Biodiversity Map. The goal, of course, is to get the entire country to light up.

We are asking landowners to remove invasive plants from their property and reduce the area they have in lawn by planting more natives. The organization's ecological products are measurable increases in biodiversity, measurable reductions in invasive plant populations, significant drawdown of atmospheric carbon dioxide, and the transformation of areas outside of parks and preserves from no-man's land into viable habitats. Our sociological products include national awareness, not just of the problems but of the solutions; recognition that nature is not optional and that everybody owns responsibility for sustaining it; a culture change from an adversarial relationship with nature to a collaborative one; and measurable progress toward the United Nations' 30×30 initiative. There is no way we will preserve ecosystem function on 30 percent of the United States by 2030 if we do not count successful conservation efforts on private property. Besides these important benefits, HNP converts hope into action. If we are successful, people will realize they can take simple steps right at home, on the farm, at their places of worship, on corporate landscapes, and at other places, that will make a difference they can see and appreciate in short order.

Do you really think homeowners can fight the extinction crisis?

Earth is now in the midst of the Holocene extinction, the sixth great extinction event in the planet's history, but this one is unique. It is not being caused by global ice sheets, great upwellings of methane from the ocean floor, massive volcanic action, or a direct hit by an asteroid. Instead, it is being caused by a single species: humans. More and more people are realizing that if humans have the power to wipe out countless species, they also have the power to conserve them, not only in parks and preserves, but in their own yards. There is no doubt in my mind that they are correct.

What's the difference between HNP and Pollinator Pathway projects?

HNP overlaps with the work of Pollinator Pathway in many ways—as it does with many other conservation groups. The non-profit Pollinator Pathway encourages people to plant pollinator gardens in enough places to create continuous strips of habitat for pollinators and other flower visitors. We like to think of HNP as the hub or the umbrella for all aligned groups, including Pollinator Pathway, amplifying what they do and providing them with resources. Our call to action is to regenerate all forms of biodiversity and thus ecosystem function on private property across the country by planting natives and removing invasives. While inclusive of pollinators, HNP is focused on restoring the entire food web. The HNP Biodiversity Map is also a key differentiator and serves as a metric of successful conservation efforts on private property, something that is not being measured in a coordinated way by any other group. We are recording the number of people registered on the map and the number of acres restored to native plant communities. That is, the map is a record of each individual's role in regenerating biodiversity. We welcome all individuals, groups, organizations, businesses, and corporations to participate and hope that they will make others aware of what they are doing. We feel an urgency to exponentially increase the number of people acting as quickly as the biodiversity crisis demands.

HNP is a non-member-based 501c3 (non-profit) that is striving for a global reach. We recently added Canada to the Biodiversity Map and expect to add other countries as well. We collaborate with many other organizations and businesses. Our singular goal is to restore biodiversity and ecosystem function on vast swaths of degraded ecosystems (which include most yards, corporate landscapes, agricultural field edges, and private woodlots). In the process, we hope to effect a culture shift whereby humans will coexist with nature, rather than banishing it to ever-shrinking parts of the globe.

Why should I get on the HNP Biodiversity Map?

You should do this for two reasons. First, recording the successful conservation happening on your property provides a quantifiable record of conservation happening on private property. Only about 12 percent of the land in the United States is federally protected. Even if we include all of the state-protected areas, we don't have enough land to prevent the biodiversity

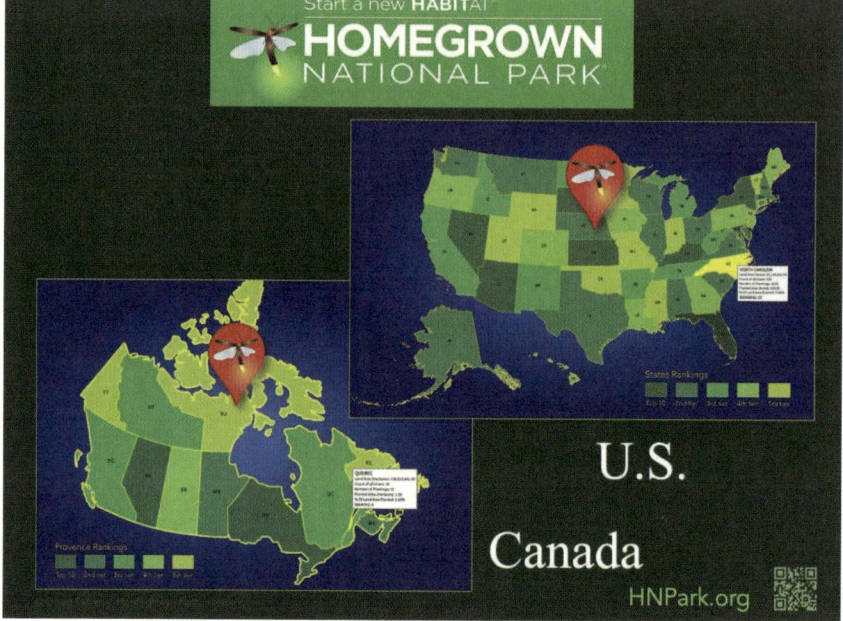

The Homegrown National Park Biodiversity Map lets us record our progress.

losses that have put us in the middle of the Holocene extinction. The solution is obvious: we have to start practicing conservation outside of parks and preserves on private property. The map provides a mechanism by which we can record our progress. The second reason we should all be on the Biodiversity Map is that it creates the opportunity for states to compete with one another. States are color-coded by the degree of participation in HNP—the greener the state, the better the participation. If you get on the map, you can help your state become the greenest in the nation!

Will HNP result in less attractive gardens?

Some gardens may be less exotic, but whether or not they are less attractive will depend on how they are designed, what plants are included, and how skillfully turf grass is used as a "social cue for care"—that is, including a little turf grass is a nod to the fact that manicured lawn is the standard-bearer of neighborhood status. There is no doubt that many exotic ornamental plants are gorgeous, and as long as they are not invasive, they too can be accents within HNP.

How do you communicate successfully with someone who is of a different mindset?

Talk from within their tribe and share personal stories that convince them you are just like them. Don't use jargon, because you will lose them right away. Remember that they are listening to you just to be polite, not because they are particularly interested. Focus on general trends, not the exceptions. Exceptions are interesting to scientists who want to learn why they are exceptions, but they muddy the waters for a novice who may be trying to understand your main points. And above all, mention "hooks" like birds or butterflies. Most people like these beautiful aspects of nature, so grab their interest with the parts of nature they already care about.

Most people don't know anything about gardening. Do you really expect them to do what HNP calls for?

Yes, I do! As "owners" of a piece of planet Earth, they also own the ethical and ecological responsibility of taking care of that piece. Not knowing how to do something does not relieve anyone of certain responsibilities. Most people don't know how to be a carpenter, or a plumber, or a roofer, but that doesn't mean they don't have to take care of their house. Most people are not tax experts, but that doesn't mean they don't have to pay taxes. In our complex society, we have division of labor: each one of us becomes an expert in a particular area, and we hire people who are experts at the things we don't know much about. If you don't know how to landscape responsibly, you should be prepared to hire someone who does.

Can the concept of HNP be extended to corporate landscapes, office parks, and industrial estates as well?

Of course! And let's not forget hospitals, schools, and houses of worship. These all add up to a lot of land. The Catholic Church, for example, is the largest landowner in the world! None of these properties is exempt from good land stewardship. Fortunately, many of these types of landowners are listening.

We're interested in developing criteria that measure an estate's contribution to local ecology. Do you have a sense of what kind of native plant surface area or percent of total land devoted to native plants, the degree of plant diversity, or the presence of certain keystone species that truly make for a meaningful impact?

No definitive answer to your question as to how much area must be converted to natives or what biomass of keystone plants is required to "make a difference" exists, because no one has funded such research yet. A study by my student Desirée Narango did show that at least 70 percent of a landscape's woody plant biomass must be native to sustain chickadee populations. Our objective with HNP is to move the needle toward that goal. A single oak tree in a yard will support caterpillars that breeding and migrating birds will use, even if that one tree is not sufficient to sustain an entire population of birds. A single milkweed clump in a container on your deck very likely will support at least one monarch caterpillar. So every little bit helps, and we don't want to discourage anyone from participating in HNP, even if their contribution is small. You can set your own minimum requirements to achieve sign status, but as an organization, we don't want to set a discouragingly high bar for anyone. Our hope is that if people try it on a small scale, they will see results and be encouraged to do more in the future.

Have you any thoughts about new housing subdivisions? What should planners be thinking about when setting the parameters for such estates in terms of aiding ecosystems?

If I had my way, landscapes in new developments would be installed with the biodiversity crisis in mind. We already have 135 million acres of residential landscapes in this country, and none were built following the four ecological pillars of sustainability: to support a complex community of pollinators, to support the local food web, to store carbon, and to manage the watershed in which they lie. When most of those landscapes were built, we weren't thinking about how humans and nature can coexist, but we know better now. All new developments should have less acreage in lawn, their dominant plants should be high-functioning contributors, and street and home outdoor lighting should emit only yellow wavelengths, with shields that direct light down instead of in all directions. While we're at it, they should install bird-friendly windows that discourage bird strikes. It is far cheaper to use bird-friendly glass when the house is being built than to retrofit after the house is built. These ecological approaches to new housing should be standard fare and would automatically qualify new landscapes to be part of HNP. If incoming residents want to rip out productive native plants and install more lawn, the cost of such deadly conversions should be on them. As it is now, homeowners who want to share their spaces with the natural world have to convert their yards at their own time and expense.

I've just gotten on the HNP map with about two acres of lawn that I will stop mowing this spring. My intention is to watch at first and see what happens. I know we'll have lots of goldenrod and milkweed. I intend to plant trees and shrubs over the years. I understand that I'll need to monitor for invasives. Do you have other suggestions or guidance?

Congratulations! I understand your instinct to stop mowing altogether, but you can speed the conversion from cool-season European grasses to warm-season native grasses by changing your mowing schedule. The first few years, mow heavily during spring until about mid-May. Then stop mowing the rest of the season. That will depress the cool-season grasses and give the warm-season grasses a leg up, and it'll also enable those milkweeds and goldenrods to colonize. You can generally follow an "addition by subtraction" approach to your restoration: monitor the area and remove any invasives as they come in, but keep the natives, which will also come in.

What have you found to be the most effective ways to reach and engage people in this project?

We do everything we possibly can to reach everyone, everywhere, to raise awareness of the crisis and solution. Homegrown National Park has an effective social media presence on Facebook, Instagram, and TikTok, where we have enjoyed more than four million views on a single TikTok post. I continue to engage in an aggressive schedule of speaking engagements both in person and online. E-news, the website, and mentions in print and online media always result in more people signing on to the map. We also collaborate with mission-aligned non-profits and groups to amplify what they are doing. The map itself is the most viable point of HNP engagement. Getting on the HNP map is the singular community-centric, interactive destination for all individuals, non-profits, businesses, and other participants, to show they are part of what is becoming the largest cooperative conservation project ever attempted. It is one of our most dynamic points of engagement.

How can land trusts be involved in HNP's goals?

Land trusts protect an enormous amount of land from development, and all of it should be registered on the HNP map to help get an accurate record of effective conservation on private property. As of January 2024, 95 land trusts are on the HNP map, protecting 32,418 acres. But we encourage *all* US land trusts to join HNP.

As volunteer master gardeners, how can we best help grow HNP?

As a master gardener, you can do a few things to help HNP. Register your property on the map and start thinking of ways you can improve the degree to which your property sustains pollinators, shares the food your plants make with animals (especially insects!), stores carbon, and manages the watershed. Planting a keystone plant would be a great start. You can also pass on HNP's message to each new class of master gardeners. Talk to your friends and neighbors who remain in the dark. Finally, you might also offer your gardening or consulting services to show people what they can do to improve their property. Life is complicated these days. In general, people are much more likely to act if someone is willing to provide some simple guidance.

I am young and don't own property, but I want to help. Can I join HNP? What do you suggest?

Volunteer! All land conservancies, parks, and preserves are underfunded and understaffed. They will appreciate any time you can provide. Help an elderly property owner who is unable to work on their property without support. And you absolutely can join HNP. Remember, it's free!

CHAPTER EIGHT

Home Landscapes

What exactly do we need to do in our yards to make them more sustainable?

Every landscape must perform four ecological functions if we are to achieve a sustainable relationship with the natural world that supports us. It's really very simple—our landscapes must do the things that enable ecosystems to produce the life support we and every other species require: they must support a diverse community of pollinators throughout the growing season, provide energy for the local food web, manage the watershed in which they lie, and remove carbon from the atmosphere where it is wreaking havoc on Earth's climate. How well a landscape accomplishes these goals depends on how well we, as landscape managers, choose and deploy the plants we include.

If we plant most or all of our property in lawn, none of these goals will be met. Lawn degrades the local watershed by discouraging infiltration, facilitating stormwater runoff, and adding nitrogen, phosphorous, herbicides, and insecticides to the nearest stream or river. Today's cultural standard for lawn supports no pollinators and does not nourish the insects that enable birds, reptiles, amphibians, and many mammals to reproduce. And when it comes to carbon capture, turf grass is the worst plant choice. Yes, lawn grasses do build their tissues out of carbon they have pulled from the atmosphere, but every time we mow the lawn, we release that carbon back into the air. Grass roots, like the roots of other plants, do leak some carbon into the soil, but grass roots are very short, and almost any other plant species deposits more carbon into surrounding soils than grass does. Finally, every time we mow, our mower belches carbon from fossil fuels into the atmosphere.

We can help our yards meet their ecological responsibilities by planting the plants that are good at supporting pollinators, good at sharing some of the energy they have harnessed from the sun with the local animals that run our ecosystems, good at holding carbon within their structures for decades or even centuries, and good at producing glomalin on root hairs, a carbon-based structure that deposits carbon in the soil. We can also choose plants with large canopies that soften the impact of pounding rain and support large root systems that encourage rainwater infiltration and thus hold tons of water on site after a storm event.

Transforming our yards into these ecological gold mines is a process that can unfold over months, years, or even decades. Every time we plant a native perennial that nourishes specialist bees, we have helped *all* local pollinators, for generalist bees can use those plants as well. Every time we add a native oak, willow, cherry, birch, cottonwood, alder, maple, or other native tree to our yard, we have reduced the ecological dead zone we call lawn and increased the ability of our yard to support breeding birds by supplying the host plants for the caterpillars on which those birds rely. Moreover, such plantings have vastly improved our yard's watershed management and carbon sequestration potential.

The path to sustainability lies along a continuum, with low (or no) ecological function at one end and a vibrant ecological machine churning out ecosystem services every minute of every day at the other. Every time we take action, our landscape moves closer to becoming a positive ecological force rather than a negative one. We have been content to make withdrawals from the ecological bank account that supports us for far too long. By helping our properties reach the four ecological goals just described, we can finally start making essential deposits.

We've been in our house in Valrico, Florida, for 30 years and have created a rich native habitat. But we have to sell, and soon we'll leave this natural oasis in a lawn chemical–soaked suburb. I hope it will pass into

good, gentle hands. As more and more native gardeners will be "aging out" of their properties, what can be done to maintain the lot-by-lot progress all of our efforts have achieved?

The idea is to change the culture to the point where people who are looking for properties will be attracted to properties like yours and will not automatically turn it back into an ecological wasteland. You should find an informed real estate agent who can guide like-minded buyers to your property. Advertise it in a way that will attract the kind of buyer you are interested in. You are not the only one who values living with nature. You also have the possibility of placing all or part of your property in a legal environmental easement. You can work with your local land conservancy to do this. Unfortunately, many worry that such an easement would lower the property value. If we had tax incentives for creating ecologically sound properties, good plantings would financially enhance property values rather than degrade them. Maybe someday.

Should I plant a woodlot with trees all the same age?

Ideally, no, but it is far better (and cheaper) to plant your trees small, which means they will all be about the same age if you plant them at the same time. You can achieve a layered landscape after just a few years, with some trees growing much taller than others, by planting species with different growth rates. For example, if you plant a tulip poplar, white oak, American holly, and white pine all on the same day, and all are 2 feet tall, then in 10 years the tulip poplar will be at least 40 feet tall, the American holly will be 7 or 8 feet tall, and the others will somewhere in between. Voila! A layered landscape!

I've considered going back to college for a degree in conservation biology or entomology so I can help save the planet, but if I can't do this, what is the next best thing I can do to help?

Work on saving your little piece of the earth. You don't need a degree to plant an oak tree, install a pollinator garden, or begin changing your homeowners association (HOA) from within. You don't need a degree to volunteer to help remove invasives in your local park or a land conservancy property. You don't need a degree to campaign against mosquito fogging or the unnecessary use of neonicotinoid insecticides. And you don't need a degree to vote for local and national politicians who understand that the environment matters.

Do native insects have issues eating or using non-native wood, or is it only the foliage they can't digest?

I don't think anyone has compared the use of natives and non-natives by wood eaters. My guess is that most of the time, native wood-eating insects won't care. Our wood eaters don't really *eat* the cellulose that is wood; it is their gut symbiotic bacteria and fungi that *digest* the cellulose for them. It is possible that some species of non-native trees such as ailanthus, autumn olive, salt cedar, and eucalyptus have distasteful secondary metabolites in their wood that would discourage wood eaters, but I'll bet most wood is not defended by toxic compounds (it is already dead, so why defend it the way leaves are defended?), but I don't know this for sure. I can tell you that mulberry takes forever to break down and may actually be an example of wood that *is* hard for insects to eat. Mulberry trees have yellow interior wood, and something nasty is making it yellow. I cut down a mulberry tree 23 years ago, and most of it is still lying where it fell, not rotted or eaten!

Management Issues

How do you handle social pressure to conform to neighborhood standards?

The other day my wife was pulling out Japanese stiltgrass from our front yard. I asked her not to weed our patch of *Bidens aristosa* (ditch daisy) because there were at least 15 goldenrod stowaway caterpillars on it; they would soon be pupating in the ground and I didn't want any to get squished by a misplaced foot. She sighed and mumbled, "This place will never look good." I replied, "Maybe not, but I guarantee we will always have the most goldenrod stowaways on the block!" And who could ask for more than that? A little snarky, I know, but the back story is that our yard is not on public display. We have a long flag lot driveway, and no one can see our yard from the street. That frees us (me) up from any public pressure (but not from spousal pressure) to

The beautiful goldenrod stowaway will not live in your yard unless its host plant, ditch daisy, is present.

focus more on yard function than aesthetics. What a luxury! But what about people who do face social pressure, which I recognize is most everybody? How can we landscape ecologically without triggering the ire of neighbors or township?

Our goal is to transition from landscapes with one goal—being pretty—to landscapes with two goals—being pretty and ecologically productive. We need to be sensitive to the fact that our landscapes reflect our values. They are an expression of status, and neatness is still and probably always will be a high-ranking cultural value.

The impression that ecologically productive landscapes are messy or neglected does not come from the plant species in those spaces. It comes from how the plants are arranged in the landscape—that is, the landscape design—and how the turf grass that is retained is maintained. The key, then, is to maintain (mow) the grass we keep and use it artfully as a "cue for care." Follow landscape architect/author Thomas Rainer's suggestion that a lawn should be more like an area rug, not wall-to-wall carpeting. Line the driveway, the sidewalk, and new beds with swaths of manicured lawn. Well-planted areas bordered by mowed turf are readily accepted by neighbors and others because it is easy to see that they are intentional designs, not the result of neglect.

I know that a large part of your work is reaching out to homeowners on regular-sized lots, but do you have any specific recommendations about managing larger properties? What things have the most impact when you are dealing with acres and not a suburban yard?

It is logical to think that larger properties pose larger restoration challenges, and in terms of controlling invasive plants and white-tailed deer, that is true. But a restoration approach that author/horticulturalist Rick Darke calls "addition by subtraction" actually makes plant acquisition easier. You simply remove what you *don't* want and leave what you *do* want. Jays, squirrels, wind, water, and seeds stored for decades in the seed bank will bring a diverse community of native plants to large tracts of land. Even those pesky deer can "plant" desirable species on occasion. (On our property, for example, persimmon trees have shown up, compliments of the deer that eat persimmon fruits elsewhere and then poop out the seeds on our property wherever the urge hits them.) Unfortunately, deer will also eat many of the seedling natives that pop up on your land, and they especially love young oaks. But they avoid autumn olive, oriental bittersweet, privet, barberry, burning bush, porcelain berry, buckthorn, and other serious invasives, which will soon dominate a property. Any resources you have for your restoration should go toward managing invasive species and deer.

One advantage large properties have over smaller suburban lots is relaxed peer pressure. Often, larger properties are not on public display like a suburban yard can be, and restoring their ecological function will not draw the ire of misguided neighbors, city councils, or HOAs. If nothing else, this means you can proceed at a more leisurely pace and employ deer cages until your plants attain sufficient height, without a second thought.

I have been removing vines and multiflora rose from a patch of land. After I cut them away, I found numerous tree saplings (4–10 feet tall) that have been growing under this brush, many of them oak trees. The issue is that they have grown bent from the pressure of the other plants on top. Their root systems seem fine. I don't really care about their appearance from the human perspective, but will they just start growing straight, now that they are unrestricted?

Years ago, I would have given you the wrong answer to this good question. But I have witnessed in my yard what happens when you plant a bent tree. Twenty-two years ago, I was gifted a 12-foot shingle oak with a main trunk shaped like an S that was, therefore, unsellable. Today that tree is 70 feet tall and its trunk is straight as an arrow. And it didn't take twenty years to straighten out, either. In just a few years, it had grown straight and tall. I wouldn't worry about your trees. They will reach for the light, and that straightens them up naturally.

In your talk, you said something I had never before heard: namely, that wood chips for mulch, although overwhelmingly preferable to bare ground or turf, can be detrimental in that they can "remove nutrients from the soil." If I heard that right, can you please explain that a little? As a commercial arborist, I certainly don't want to deplete any soil.

You hear correctly: woodchips do reduce soil nitrogen in the process of decomposing. I was also surprised the first time I heard about nitrogen depletion by wood chips. The information came from the greenhouse manager at the University of Delaware back in the 1980s, but I have heard it several times since. It seems the microbial activity required to decompose woodchips is so intense that it draws nitrogen out of the top few inches of soil to support the growth of the bacteria and fungi in the woodchips.

I am on the grounds crew at a local university. A coworker suggested that I get rid of a staghorn sumac, but I told him that bluebirds like the fruit. Are there any other reasons to keep it around?

Please do not get rid of your staghorn sumac. In fact, plant more! It makes great berries for lots of birds, not just bluebirds, and it is the host plant for the showy emerald moth. It's also a great soil stabilizer. Any grassy slope should be replaced with staghorn or smooth sumac.

We have about a 3-to-1 slope on half of our property, and it is prone to erosion that I would like to stop with plants. I would prefer to plant trees on the slope without leveling a section and then use a deep watering system in the early years to help the tree get water. Trees seem to grow on slopes in nature, but most sources say you need to level off an area in the slope for a tree to survive. Is that your experience as well?

The showy emerald moth is just one of the treasures supported by native sumac species.

No way! Trees grow on slopes in nature all the time, and no one levels the area to get them started. That's because they start from seed in nature. So the key is to plant them small. That way, it will be easy to get them enough water until they are established. The reason a big tree transplant needs lots of water is because most of the roots were cut off to move the tree. So you have to water constantly to compensate for root loss. When you plant a small tree (3 feet tall or less), the root-to-leaf ratio is much closer to 1 to 1, and this means that it has plenty of roots to take up water for the number of leaves that need it. I have planted trees on slopes many times. I dig a hole, plant the tree, and then build a small berm on the down slope, so when I water it after planting, the water can seep in rather than run off.

One of our oaks was hit by lightning. We recently saw a program about woodpeckers and how they love dead trees. Our backyard is a woodpecker's wonderland, with Pileated, Downy, Hairy, and Red-bellied woodpeckers, and Northern Flickers. Our oak looks like it will die. Can we save part of the trunk for the woodpeckers?

Pileated Woodpeckers are just one of 87 cavity nesters in North America that benefit from snags in our landscapes.

Absolutely! Cut off the bulk of the big limbs—those that may fall when dead. Then top the tree as high up as you feel comfortable. It will stand as a dead snag for decades and provide food and homes for your woodpeckers. Homeowners are having great success with this approach in Texas with large loblolly pines. Front yards, backyards—it is working well wherever they do it. Brown-headed Nuthatches can excavate a nest hole within a year!

I am removing sod around my hilltop home. Sod will be replaced with keystone type natives. Erosion will be an issue. How do you feel about glyphosate use? So far, I have avoided it using a sod cutter, erosion-control blankets, and heavy planting in full sun exposure, but some remaining areas are tricky, with five mature oaks (lucky me). But it just seems wrong to use the herbicide.

I avoid herbicides when I can, but there are times when they are the best approach. Removing grass can be one of those times. It is difficult to remove sod without removing a lot of topsoil. You can pile lots of leaves on your grass to smother it or cover it with heavy black plastic to cook it, but those approaches are time consuming and can take months or even years to work. Herbicides are an important tool in our management toolbox; they can be misused and overused, but when properly used, they facilitate the removal of turf grass and invasive plants. I consider them a form of chemotherapy. The chemicals we use to fight cancer in our bodies are nasty but necessary—but, of course, they have to be used properly or they kill the patient. In my view, the plusses that result from removing lawn far outweigh the minuses of proper and limited herbicide use. If you decide to use a herbicide, however, don't choose a product labeled "extended control" or "season-long." These products will keep killing far longer than necessary, especially since you want to replant the area. In addition, avoid spraying when rain is expected within 48 hours to avoid any herbicide washing into nearby waterways.

What is a good starter book or online resource for beginners to learn the proper techniques for planting and maintaining trees?

Dan Lambe and Lorene Edwards Forkner's book *Now Is the Time for Trees* covers the proper way to plant a tree. A more focused option is Catherine

Copp's book, *Mighty Oaks from Little Acorns: The Complete Guide to Growing Oak Trees from Seed*. My general guidelines are simple: First, plant as young a tree as possible, and avoid disturbing its roots. It will grow faster and be healthier than a larger tree that needs its roots pruned for transplanting. Second, do not dig your hole too deep. A shallow, wide hole is far better than a narrow, deep hole. You don't want the tree to settle after planting so that the soil line is above the root crown, because that will smother the tree. Don't backfill the hole with amended soil. Amended soil usually settles quite a bit, and your tree will drop in its hole along with the settled soil. Finally, do not pile mulch against the trunk of the tree, as if it were a volcano. That is a sure way to introduce trunk rot.

Is companion planting with species such as garlic and marigolds that repel insects still recommended for managing vegetable garden pests?

Decades ago, Richard Root at Cornell University advanced what he called the "resource concentration hypothesis." In a nutshell, it proposed that when you plant a group of a single plant species in one place, insects that use that plant are more likely to find and exploit it. But if you plant a species amid many other plant species, by interplanting, insects will have a more difficult time locating it among the other species.

Studies of cabbage white butterfly larvae (cabbage worms) have suggested that Dr. Root was right. Cabbage whites are introduced insects with few natural enemies; thus, their pest status. Cabbage worms feed only on host species of cabbage, broccoli, kale, and other brassicas; they cannot eat other plants. To keep cabbage whites at bay, you can "hide" the brassicas by interplanting them with strong-smelling deterrent plants (such as garlic, dill, fennel, basil, allium, thyme, tomatoes, onions, sage, borage, nasturtium, tansy, and rosemary). Plant the deterrent species first and allow them to establish before interplanting the brassicas; adding deterrent plants to an established cabbage patch that already has cabbage worms won't help get rid of the pests.

Can butterfly host trees such as black cherry, hoptree, and hackberry be coppiced or otherwise trimmed to keep them from outgrowing a small yard space?

Yes, indeed they can, and I wish more people would consider this compromise. Many species of broadleaf trees respond well to coppicing, with oak, ash, alder, sycamore, hazelnut, and willow leading the way. Black cherry is not supposed to coppice as well, but I have not found that to be the case in my yard. Fall is a great time to coppice. When your tree reaches a diameter of at least 2–3 inches, cut it off about 12 inches from the ground. Your stump, or stool, will sprout new growth the following spring. When it does, you have created a copse, which can be cut again and again, indefinitely.

A great way to fit a large tree species into a small property is to coppice a young tree, like this Oregon oak, into a shrub.

I am trying to make my yard friendly to all. Can I just ignore the voles, or must I try and control them?

Well, I ignore the voles in my yard, and the foxes, hawks, and black rat snakes appreciate that I do. Occasionally, I grit my teeth when a vole girdles a young beech during winter, but life is full of trade-offs. Voles are important components of local food webs, and I wouldn't be able to enjoy other creatures or the ecosystem balance they provide without voles. Please tolerate your voles if you can.

Recently, a tornado cut through our property and it topped, and in some cases uprooted, a number of oaks. We own 135 acres of mixed hardwoods, with many oaks. Is there anything we should be concerned about now? Like disease?

Catastrophic disturbances like tornados do change the status quo, but they also open up new niches for plants and animals that may have been excluded from your site due to the lack of light. They also create snags for cavity-nesting birds and add coarse, woody debris on the ground for countless ground dwellers. I wouldn't worry about disease, but I would worry about white-tailed deer that will eat all the native plants that germinate in the disturbed area while not touching the invasives that will inevitably come in, leaving you with a tangle of invasive vines and shrubs that will support little. I would advise that you cage seedlings of the canopy trees that sprout and spot treat the invasives as they appear. The disturbed areas will produce lots of young growth and productive edge habitat in just a few years.

Should I cut Virginia creeper off my trees?

No! Virginia creeper is a wonderful native plant. It can climb trees without girdling them, and it is lightweight so it does not pull them down. It's also a great pollinator plant; its flowers are small and inconspicuous, but you will know when it's in flower because of the cloud of native bees it attracts. In autumn, those pollinated flowers turn into nutritious berries that are 25 percent fat, exactly what migrating and resident birds need for food in fall. Virginia creeper is also a great host plant for the large sphinx moth caterpillars that are important food sources for nestling cardinals—species such as Pandora

Virginia creeper is a productive native vine with beautiful fall color.

sphinx, lettered sphinx, hog sphinx, and Abbott's sphinx. Finally, Virginia creeper is a great groundcover and has excellent fall color, a plus all around.

Is goat management of undesired vegetation advisable?

There is nothing wrong with goat control of invasives except that it is expensive and works only temporarily. Goats will eat everything back to the roots (including desirable natives), but they do not kill the roots. So when the goats leave, the invasives roar back. One could argue that they are good for opening an invaded area so that you can easily gain access and actually kill invasive root stocks, but goats will not permanently solve an invasives problem.

As we document the effects of climate change more and more, I often think about how some more southern species may become increasingly important for wildlife. Wouldn't the trees that moved north fastest also be the fastest to adapt? Many environmental professionals still want locally native species and ecotypes rather than more southern species of natives. Could you provide your thoughts on the use of southern, more heat-tolerant species within more northern parts of the United States?

You are talking about assisted migration, a controversial plan to manage plant populations in the age of climate change. There are two reasons I'm not a big fan, in most cases. First, climate change is not causing a gentle rise in temperatures, as the term *global warming* implies. We are consistently in the record books for wild winter weather (temperature drops of 40 degrees in a half hour in Denver, for just one example). That's rough stuff, not just for animals, but for plants as well. So a more accurate description of what's happening might be "drastic climate variability." Yes, the average temperature around the globe is going up, but what is more noticeable and destructive are the wild swings in temperatures we are seeing quite often these days. These wild swings are the result of a "poorly behaved" jet stream that is wobbling much more than it used to. When it wobbles north, we experience temperatures much warmer than "normal," but when it dives south, it drags very cold arctic air along with it. If we move plants adapted to southern climates north, they will do well during the warm spells but will die during the cold extremes caused by southern wobbles.

Will the trees that moved north the fastest also be the fastest to adapt? The short answer is no. Remember what adaptation (evolution) is: a change in gene frequency within a population. No organism adapts genetically to an environmental challenge in a single generation. The genes you are born with are the genes you die with. In other words, it takes several generations, usually dozens, for natural selection to weed out the individuals that are not adapted to local conditions and favor the individuals that are. Moving southern populations of trees north would expose them to northern conditions first, but the tone of your question suggested that you thought those individuals would start to adapt immediately. If the trees are not killed outright by a polar vortex (a distinct possibility), natural selection will start to act on their offspring and their offspring's offspring.

The second reason I don't support assisted migration is this: if we take a southern plant out of the community of plants and animals it coevolved with and move it to a more northern community, we will leave behind the species that have interacted with that plant for eons, including the insect specialists that depend on that particular plant for their existence. As is the case whenever we move a plant beyond its natural range, moving southern plants north makes them less productive members of their new ecosystem.

There may be a case here and there where assisted migration makes sense, particularly if we are moving a plant just a bit more north of its current northern limit, but in general, I think it makes more sense to have faith in the existing genetic variability in plant populations. Genetic variation enables plants to adapt to environmental changes, including the climate challenges we have inflicted upon them.

I live on four acres in southeastern Massachusetts in an oak/pine forest where the white pines are taking over. To give the oak saplings more light, I've been removing pine saplings around as many oaks as I can. Is this helpful?

Well, it is helpful if you want to manage an oak woodland, but it sounds like you might be forcing oaks into a forest that wants to be white pine. A few centuries ago, portions of Massachusetts were dominated by mature white pine forests with enormous trees, 250 feet tall. We logged them all out, and I do mean *all* of them, and the logged areas were subsequently colonized by other species, including oaks. Oaks are wonderful trees, but so are white pines. Of course, a monoculture of one species is less desirable than a forest with more diversity, but don't fight the special qualities of your site. If white pines do better there than oaks, go with that. If I were you, I would clear around a few of the oaks and let the rest go to pines.

A certified prescribed fire burner I know tends to think that fire is *the* answer to producing biodiversity. I am afraid of using fire except as one tool in the box. Any thoughts?

Fire is indeed an important tool in our management toolbox that is good for meadows/prairies and all of the plants and creatures they sustain. But like any tool, it has to be used properly. The most common mistake is to burn too often and too thoroughly. Even in the prairie fire climax communities of the Midwest, fire occurred naturally only once every three or four years, and when it did occur, it burned patchily. This is an important point, because the patches that *don't* burn provide refuge for the insects that are needed to recolonize the areas that *do* burn. Unfortunately, fire is ineffective in controlling woody invasive plants. It doesn't kill their roots, so they come right back after a burn. If you have someone willing to burn for free, take advantage. It is expensive to hire a certified crew. I'm not sure whether you are afraid of the danger (don't be—a trained burn crew can start and stop a fire on a dime) or because burning will kill your trees. If you are trying to reforest an area, burning is not the management tool of choice. That said, because of European colonists' suppression of fire in the eastern United States, competition from black cherries, birches, tulip trees, and other deciduous species has caused our oak densities to fall by 50 percent in the last 150 years. So fire is an important tool in the management of oak forests as well as meadows and prairies.

Doesn't a well-planted landscape put you at risk from fires, at least in the West?

The risk of fire in the West must be taken seriously, but denuding the landscape wherever we humans go to minimize fire risk is not an option because it would cause an ecological disaster as big and as far-reaching as the megafires we are experiencing. It is possible, however, to design landscapes that minimize fire risk.

Before we started to mismanage our coniferous forests in the West by suppressing fire and allowing fuel to accumulate, frequent low, relatively cool ground fires were occurring. They would consume accumulated grasses and shrubs every few years without jumping to the crowns of the large trees. We can re-create this savanna-like landscape by thinning the large trees near our homes and removing dead brush every few years just as a ground fire would do. Experts, such as Cassy Aoyagi, who specialize in fire-resistant landscapes suggest we do this in a 100-foot circumference around the house. In California, the use of live oaks, such as valley oak, blue oak, and coast live oak, provides a fire break when canopy fires send burning embers hurtling through the air for hundreds of yards. Live oaks intercept those embers and snuff them out before they reach our highly flammable buildings. In fact, studies have shown that our built structures actually fuel these fires as much or more than vegetation.

Another tip is to be sure to remove invasive cheatgrass (*Bromus tectorum*), a winter annual from Europe that has spread throughout the West. Though green in winter and spring, cheatgrass dies back to highly flammable tinder in early summer. Most terrible forest fires actually start when cheatgrass is ignited by lightning or careless people. Eradication of cheatgrass is unlikely, but you can suppress its dominance by encouraging native perennials that hold more moisture through the dry summer months and are thus less flammable.

Our cabin in northwest Wisconsin is surrounded by many oaks, mostly red oak, with a few hickories and birches interspersed (also a few evergreens and one maple). Are there any interrelationships among these trees that I should nurture, or can I clear a few of the species I like least (hickory) to create more light for the rest to grow stronger?

All of the trees you mention are highly productive. Your oaks are great, but so are those hickories you don't like, as well as the birches and maples. Depending on the soil type, oak-hickory forests are a natural phenomenon—that is, these two species "like each other." If I were you, I would work to retain all of those species as well as on letting in a little more light. That's a challenge that will keep you busy for a while!

What are your views on spraying turf for broadleaf weeds? My university now has contracted to spray its large swaths of turf areas. I wonder what I could tell my workmates, sustainability committee, or president of my school to prevent this? From your books, I learn the importance of violets for pollinators, but could you tell me whether dandelion, plantain, and other weeds are beneficial?

I am totally against spraying grass to kill broadleaf species. The whole exercise is strictly for aesthetics, and much of the herbicide gets into the watershed. White clover and dandelions are non-natives, but they do supply nectar for generalist pollinators and honey bees. Violets are native and so are several species of plantain; both feed several species of caterpillars. But even if they fed nothing, polluting our watershed so that we can have 44 million acres of perfect lawn is only helping the herbicide companies. It simply isn't worth the horrendous ecological cost. Mowing keeps those "weeds" at bay just fine.

One issue that I've never seen addressed relates to septic system mounds. We were told not to plant anything on our septic mound, so what we have is a large swath of lawn. Do you have any recommendations for beneficial plants that would not interfere with the mound system's purpose?

Oooh. I hope an inspector never sees my septic mound! Enough said. Septic mounds are great places to install a meadow/prairie. Meadow and prairie roots can grow fairly deep, but they are wispy little things that will not wreck your concrete-lined septic system. It is so much more productive to put a diversity of pollinator plants over your septic system than to go simply with lawn.

Do I have to get rid of my lawn before I plant my natives?

That depends on what natives you are planting. If you are adding trees, the answer is no. You can plant right through your lawn, although making a lawn-free bed under your new tree, even if it is just 3 inches tall, is highly recommended. That is a great way to reduce the area of lawn in your yard

Meadow plants can be seeded directly over a septic field without causing harm.

and to create a soft landing for the caterpillars that will develop on your tree. Your new trees will grow faster than you think and justify that new bed before you know it.

If you are planting a pocket prairie or a bigger meadow/prairie, killing the turf grass before seeding or adding plugs is strongly advised. Grass will present a weeding challenge for years after planting if you don't get rid of it before planting. I do not recommend digging up your turf; it is very difficult to remove grass by digging without taking lots of valuable topsoil with it. Using a smothering technique, such as black plastic or builder's paper, or spraying with glyphosate are better options. Great ways to convert lawn to meadow are discussed in Neil Diboll and Hilary Cox's 2023 book, *The Gardener's Guide to Prairie Plants*.

Should I focus on making a pollinator garden or should I focus on caterpillars?

Ideally you would do both. We need both caterpillars and pollinators, particularly specialist bees. The amount of available sun may dictate which approach you favor, if you have to choose. If your yard is already heavily shaded, you probably don't need to add a keystone tree species, unless your shade is coming from Norway maples or some other non-natives. But you could add a patch of goldenrod or native asters in a sunny spot, even if it's a small patch. And don't forget the container gardening option. Mobile containers planted with flowering natives can be placed in any area with a bit of sun. If you have the opposite situation—all sun and few to no trees—you may have plenty of room to create a diverse pollinator garden and add an oak or two. It will be years before that oak shades out your pollinator garden. In the meantime, you will be making caterpillars *and* pollinators, the best of both worlds.

How can I manage the woodies in my meadow by mowing without hurting the biodiversity in the meadow?

Unfortunately, mowing is only a temporary (stopgap) approach to controlling woody plants in a meadow. That is, it doesn't really control them because mowing doesn't kill their roots. Spot treatment of woodies, either by whacking them out with a mattock as they appear each year or by treating with an herbicide, will be required at some point. But let's say you are going to mow in the meantime. The general consensus for proper mowing or burning is to avoid mowing or burning a particular spot more than once every three years. Mowing or burning kills the animal life where it occurs; therefore, in any given year, the recommendation is to mow or burn only a third of your meadow. That way, the two-thirds you don't mow will provide colonizers for the third you do mow, and your meadow will flourish continuously. If you mow or burn the entire meadow at one time, the new growth will have to be colonized from outside of your system, and it could take years to reassemble the entire complement of species that belong there.

I am the mayor of a small town. What can I do to further your message?

I am so glad you asked, because there is much you can do to change the landscaping culture in your township. You can use your position to influence new plantings on your town's public property and never permit the planting of invasive species like Callery pear, burning bush, barberry, privet, bamboo, eucalyptus, Brazilian peppertree, and so many more. As funding permits, request that township buildings be invasive-free and dominated by natives; they serve as model landscapes for residents and should be landscaped in ecologically responsible ways. Reduce roadside mowing to once a year in March—more than that is unnecessary, adds pollutants to the atmosphere, and prevents the establishment of pollinator and monarch roadside habitat. Consider replacing the white lightbulbs on outdoor township buildings with yellow bulbs that do not attract and kill nocturnal insects. Prohibit mosquito fogging; despite what the fogging companies say, fogging kills all insects but does not control mosquitoes. Switch to mosquito dunks that *do* control mosquitoes. Finally, eliminate outdated township weed ordinances, which are antithetical to good landscape management. Let me know when you finish all of that, and I'll nominate you for mayor of the year!

Can you explain the importance of improving soil quality?

Amending your soil is less important than you think, particularly when you choose the right native plant for the right place. Adding organic matter to soil in the form of leaf mulch is nearly always a good thing, but many native North American plants, particularly those in areas that were once covered by glaciers, are adapted to the poor soil conditions. Butterflyweed, *Asclepias tuberosa*, provides a great example. This beautiful native is a great host plant for monarch butterfly larvae and is very popular in the native plant trade. People buy it, take it home, and plant it the way they would most ornamental plants, adding mulch and fertilizer to the planting hole. Two weeks later, they find the plant has died. Butterflyweed is just one of many species that thrives in poor soils and dies in rich soils. One thing you should always do, however, is loosen compacted soils (as much of our urban soils are) using a garden fork before planting.

How do I decide when to remove a tree?

Before removing a tree, seek advice from someone other than the person who will make money if you decide to remove it. Most people remove trees that threaten their house if they were to fall in a storm, but taking down a tree because one of its limbs has died or it has a hollow space in the trunk is usually unnecessary. Those are common features of healthy trees and do not spell its doom.

How do I remove large non-native trees from my yard and add small native trees without wrecking the watershed during the transition?

Proceed slowly. You do not need to remove all of your non-natives on the same day. Removing one non-native a year if they are large trees seems reasonable to me. That will give the small replacements time to establish their root systems and will help manage the watershed.

Does the European beech nurture insects, birds, and mammals almost as well as the American beech?

No—just as Chinese chestnut cannot take the place of American chestnut in our food webs, Chinese ash can't functionally replace our North American ash species, sawtooth oak from Asia can't replace our native oaks, English oak from Europe can't equal the value of our native oaks, and Norway maple can't replace native maples. Even though these non-native plants are relatives of our native congeners (plants that belong in the same genus), they support on average only 32 percent of the insects that their native counterparts support.

Are hostas ecologically good ornamentals?

They are pretty, the deer love them, and they are not invasive. They don't support any insects save a few generalist pollinators, so they don't add energy to the food web, but they could easily fit into your landscape without ruining the food web as long as native plants that *do* contribute energy to the food web dominate the landscape.

Do you recommend removing non-native ornamentals such as Japanese maples and Asian cherry trees from our yards?

I recommend creating landscapes that are dominated by productive native plants. But that doesn't mean there is no room for attractive ornamentals such as Japanese maple, Asian cherries, camellias, forsythia, or any of the other non-native ornamentals that are not serious invasives. I was chastised the other day when I talked about such compromises in plant choice because, my chastiser said, we never know when a commonly used ornamental will become invasive. And there is some truth to that. But as time goes on and certain species continue to behave themselves, it becomes less and less likely that they will suddenly become invasive. All of the plants I've mentioned have been employed as ornamentals in the United States for well over a century without exhibiting the traits that make plants invasives. So if your yard is populated only with non-native ornamentals, you can indeed remove some and replace them with oaks, birches, hickories, and other natives. But my guess is you can reach the native dominance I recommend by adding keystone natives without removing anything but some lawn.

What do you think about horticulturist/author Jessica Walliser's 50-degree rule, which says to delay spring clean-up until temperatures are consistently in the 50s? I've seen varying day counts: 5 consecutive days over 50 degrees, 7–10 days above 50, and so on. Do they need to be in a row? Could you give us more information on when certain groups of insects start to emerge (native bees in hollow stems or nesting under grasses, beetles in leaf litter, caterpillars in cocoons/chrysalis, and so on)?

Insects do become active based on a combination of day length and accumulated degree days—that is, there must be a certain number of days above a given temperature for the insect to complete its development within its chrysalis or pupa so it can emerge as an adult. There are also several species that overwinter as adults, as larvae, or as eggs, whose activity is triggered by warm temperatures and lengthening days. The problem is that each species requires its own specific temperatures to become active, and one prediction of some number of days above 50 degrees does not pertain to all insects. Species of moths, butterflies, bees, beetles, and other insects emerge all season long—some in spring, some in summer, and some in fall. For example, at our house in southeast Pennsylvania, beautiful luna moths and io moths do not emerge from their cocoons nestled within leaf litter until early May; various species of oak leafminers don't emerge as adults until mid-July, and the velvetbean moth doesn't appear until the end of August at the earliest.

The reason insects appear at different times during spring, summer, and fall is that the food their larvae require isn't ready all at the same time. The evening primrose moth, for example, doesn't emerge as an adult until late July because its larvae develop on seed pods of evening primrose, and seed pods aren't produced until early August. A number of moth species are called winter moths because they don't emerge as adults until October; they then remain active on warmer nights all winter long. The same story of staggered emergence can be told for native bee species: some are active for a few weeks in March, others from June through August, and others are most active in September. So you really can't pulverize all of your leaf litter or cut down the stalks of all of last year's meadow plants after a few days at 50 degrees in spring without hurting some of the creatures within.

How does what you and Homegrown National Park are promoting fit into regenerative agriculture?

Simply put, regenerative agriculture is an approach to food and fiber production that is focused on ecological sustainability as much as product. It favors natural processes that minimize human intervention. Instead of using chemical fertilizers or pesticides, farmers employing regenerative agriculture use composting, crop rotation, prairie strips, and companion planting, and they always try to minimize soil erosion and restore biodiversity while they produce crops and livestock.

My message of including more natives in our landscapes is entirely consistent with regenerative agriculture. Intentionally incorporating native trees and shrubs into the agricultural production systems adds species diversity, habitat, and multiple ecosystem services at the field level, particularly to rangeland watercourses. Managing verge/border spaces in agricultural landscapes with native flowers, shrubs, and trees to add the plant diversity necessary to support native bees, monarchs, birds, and natural enemies, in my mind, fits under the regenerative agriculture philosophy as well. Regenerative agriculture could be one of the most powerful tools in our conservation toolbox, because so much land, nearly one billion acres—or 50 percent of the United States—is allocated to one form of agriculture or another.

Why do some trees blow over so easily?

Nearly any tree species is vulnerable to blowdown when it is planted as an isolated specimen, unable to interlock its roots with those of neighboring trees. When we get a lot of rain followed by wind, the soggy soil is not enough to hold top-heavy specimen trees upright. We can minimize residential blowdowns by planting small tree groves: groups of two or more trees spaced as they would appear in a forest setting. Trees in these small groupings can interlock their roots to form a very stable matrix far less susceptible to blowdowns.

Do landscapes dominated by native plants have to be messy?

No, they do not. Formality (lack of messiness) is a function of the garden design, not the plants in the garden. North American native plants are used in the formal gardens of Europe every day. And the non-native ornamentals we love so much in this country grow wild and untamed in their native lands. How we use these plants determines whether we have "messy" landscapes. Professor Shaun McNiff of Lesley University tells us that "When order is perceived in the environment, there is a corresponding feeling of order within the mind and body of the perceiver." I don't know what order in my mind and body feels like, but I do know that we humans feel safer when we live in open landscapes with no space for enemies to hide. I maintain, though, that with the proper design and liberal use of "cues for care" such as neat borders and swaths of mowed turf, we can have beautiful, ecologically productive landscapes that do not trigger our ancestral fears.

One hundred percent of this formal garden at Mt. Cuba Center in Hockessin, Delaware, is made up of native plants.

How can we make cemeteries more ecologically friendly?

Add more native trees wherever possible and create beds with shrubs and groundcovers rather than grass under those trees.

One of our gold-medal gardens in New Orleans just got cited for weeds and rodent harboring. I was wondering how best to approach the New Orleans Department of Parks and Parkways, plus the local government, to help with their laws concerning the definition of "blight."

Here are a few tips. First, make them stick to the facts: get them to show you the study that shows more rodents (meaning rats) come into a garden that includes native plantings. They can't, because there are no such studies. Rats are associated with human garbage, not native plants. You can also tell them there is now a legal precedent favoring native plantings in residential landscapes.

What they really fear is not rats and blight. It is lowered property values. This is where the judicial use of lawns comes into play. I do not suggest getting rid of lawn; I suggest reducing the area that is in lawn and using it as a "cue for care." Edge sidewalks, driveways, and new beds with a mower's width of manicured lawn; this shows that you understand the culture, you really are a good citizen, and your plant-packed yard will not lower property values. Remember that we elect community leaders. They work for us; we don't work for them. So if an issue is important to us, we need to make it known to our elected officials. They will listen if they think their jobs are in jeopardy.

When is the best time to clean up the wildflower garden after winter? We live in central Florida, and winter, such as it is here, has come and we have many brown stems that need to be taken down. I know we should wait, but for how long?

You won't like this answer, but many species are using those brown stems as overwintering sites or will use them as nesting sites the following summer. Conservationist Heather Holm recommends that you cut off those stems 12–15 inches above the ground. This somewhat reduces their unsightliness, yet still preserves valuable nesting sites for native bees in the coming spring and summer. Unfortunately, cutting off the tops of those stems removes the seeds that overwintering birds depend on for food, so the best time to take them down is early spring after most of the seeds have been eaten. Even then, if you can stack them out of sight, it will enable the many eggs from field katydids to hatch the following summer.

Giving up non-native vegetables seems like a nonstarter for most people. Before reading your work, I never had reason to deeply consider the origins of common vegetables. How does planting non-native vegetables, such as broccoli, carrots, and cauliflower, effect the ecosystem?

Oh my! I am not asking you to give up non-native vegetables! If we gave up growing all vegetables that weren't native to North America proper, we would have no tomatoes, potatoes, eggplant, lettuce, peppers, squashes, carrots, sweet potatoes, peanuts, spinach, corn, or beans. My request to increase the amount of native plants on our landscapes is not directed at vegetable growers. Even if all of our agricultural plants were native, we are growing them for our own food, not to share with wildlife. And there are many positive aspects to growing vegetables at home—for example, you can control what chemicals, if any, are applied to those plants, no transportation costs are involved in getting that food to your table, and so on. My message is focused on the 52 percent of the land in this country that is not in some form of agriculture. That said, most vegetable gardens *do* have enough space for a milkweed or goldenrod patch or two.

Can solar farms be used to help biodiversity?

Solar farms can be great opportunities for helping biodiversity. Solar panels capture the sun's energy directly and convert it to electricity. Plants do nearly the same thing: they convert energy from the sun into the food that supports the earth's animals. Doing both on the same piece of land is a win-win!

There has been a lot of research regarding what plants are best at supporting the caterpillars that drive most terrestrial food webs. And we are learning more and more about the best plants for pollinators. If you combine those two areas of knowledge, you can create a very productive landscape. As far as caterpillars go, the most productive plants are trees, and most would be

Solar arrays like this one at the University of Illinois, can support productive pollinator gardens.

too large for solar farms. But willows are ranked right near the top, and there are many species of short scrub willows that would be perfect for borders and maybe even within the farm itself, as well as native viburnums and ninebark. *Solidago* species (goldenrods) are top-ranked perennials for both bees and caterpillars. Native asters, perennial sunflowers, violets, and evening primrose are also highly ranked.

One important caveat needs to be mentioned. I am hearing more and more about mature forests (one project in Rhode Island and one in Virginia come to mind) being cut down to build solar farms. This is beyond ridiculous and should never be permitted. It's like draining a large reservoir in order to build a desalinization plant. Forests *are* solar farms, and though they don't generate electricity, they do draw down huge amounts of atmospheric carbon and store it in their tissues and the soil beneath them for thousands of years. They also support pollinators, manage watersheds, remove pollutants from the air, moderate weather extremes, generate food webs that support the animals that help run the ecosystems that support us, and more. It is high time—no, way *past* time—that we treated our forests as natural resources essential for the common good and protect them with that urgency in mind.

How can I maintain biodiversity in a medium-sized, one- or two-acre pollinator planting? It seems that the diversity of the native plants declines, reverting to just a few aggressive species such as Canada goldenrod, tall boneset, cup plants, and switchgrass about 10–15 years after installation. How can we manage to prevent that?

The big missing factor in maintaining diverse grassland/prairie/meadow restorations is grazing. These ecosystems coevolved with large mammalian grazers for many millions of years, and the grazers selected for grasses in the first place. By always eating the tops of plants, grazers selected for the evolution of meristematic tissue (the cells that produce new plant material) down near the ground instead of at the plant tops. That's why we can mow grass without killing it. We never mow the meristem. When we remove grazers from the system, fast-growing dominant species such as switchgrass or Canada goldenrod that are controlled by grazing tend to push out many of the other grasses and forbs. The answer to maintaining meadow diversity is to bring grazers back periodically (a few cows, horses, or even brief visits by sheep) or simulate grazing by mowing or selectively weed whacking. I know, that's a pain in the neck, but we can't expect nature to function the way it evolved to function when we remove half of its interdependent parts.

How should I handle heavy clay soil?

Don't fight it. Don't use soil amendments. All you need to do is choose regionally appropriate plants adapted to bottomland conditions. Plant perennials such as coneflower, beebalm, native aster, black-eyed Susan, big bluestem, goldenrod, cardinal flower, goat's beard, many milkweeds, tickseed, and blazing star. Plant shrubs like buttonbush, winterberry, inkberry, sweet pepperbush, and wild indigo. And plant trees like bald cypress, river birch, maple, hackberry, willow, swamp white oak, bur oak, swamp chestnut oak, pin oak, green ash, alder, bald cypress (in the South), catalpa, shadbush, sweetgum, native crabapple, sycamore, native elm, poplar, American hornbeam, and many others.

What should I do about existing landscape fabric? Should I remove it or ignore it?

I recommend removing landscape fabric, especially when it is made of plastic. It is designed to inhibit root growth and it does so forever. Landscape fabric is a lazy man's solution to weeding. A better solution is to plant so heavily that weeds have trouble getting a foothold. Green mulch (groundcovers) prevents weeds, while other mulches encourage them.

The best way to minimize weeding is to plant dense groundcovers that naturally keep out the weeds.

In our attempt to manage our woodlot, we wonder about the best solution for dead trees that are leaners. Are we (and the birds and insects) better off leaving the leaners in place or bringing them down to the ground (for insects or for our firewood)? For the big picture, what is the best way for us to learn about managing our small bit of woods?

Leaners are halfway between valuable wildlife snags and coarse woody debris when all the trunks and branches have fallen to the ground. Both play important ecological roles on your property, so keeping as much around as you can would be best for wildlife. Like you, I heat with wood that was grown on our 10 acres, so I understand the need for compromise between what to take and what to leave. My gut reaction is to take your leaners before you take ground wood, and think about creating a more savanna-like landscape through thinning, while maintaining a varied age structure of trees at all times (seedlings, understory trees, as well as those big canopy trees). You need young ones to replace the big guys when they go. As for excellent general tips on how to manage a woodlot, check out the book *The Woods in Your Backyard*, co-authored by Jonathan Kays.

Urban/Homeowners Association Issues

Do you have a short and to-the-point statement to submit to my HOA, which recently cited me for "growing weeds"? Those "weeds" included prairie dropseed, liatris, agastache and salvia.

How about this: We humans face life-threatening crises from climate change and biodiversity loss. Reducing the area of lawn and using more productive native plants in our landscapes addresses both of these crises. This can be done tastefully without the use of chemicals and without reducing property values. It is the future of landscaping.

How can I convince my HOA to landscape more sustainably?

Infiltrate and educate your HOA. Change from within. That is so much easier than contentious and costly lawsuits. Provide a great example of the type of landscaping you are suggesting so that it can be a reference point for all involved. I am getting more and more emails describing how well this works. Here is one, edited and abbreviated:

> I proposed that my condo association adopt a sustainable landscaping policy, to be followed by the creation of a landscaping plan with steps and goals to achieve the policy. Our board of directors voted to adopt the policy, which reads: "The condo association commits to following sustainable landscaping practices aimed at improving air and water quality, conserving resources, limiting the use of pesticides, and supporting wildlife habitat through the management of invasive plants and the increased use of native species, with the eventual goal of having native plants comprise at least 70 percent of our community's plant biomass." This is a long-term goal, since we can't afford to replace our many healthy non-native plants, and we need to try to win over residents who think native plants are messy and who selected many of the current ornamental plants, but it's a big step in the right direction. I joined our association's landscaping committee to try to achieve this. Last year I ran for election to our board in the hopes of being able to bring about change more effectively, and now we have our sustainable landscaping policy.

How can a non-scientist estimate the biomass of plants? My HOA has adopted a new sustainable landscaping policy with an eventual goal of achieving a 70 percent biomass of native plants, in line with your research. But I can't find a good way to do this, since we won't be tearing out plants, drying them, and then weighing them.

Estimating plant biomass doesn't need to be that accurate or that difficult. What you are really doing is estimating the volume of leaves produced by a given plant. Get a piece of graph paper and roughly draw shapes representing the areas occupied by the leaves of each plant species in your yard. The graph paper will help you estimate relative size. Label each shape, cut them out, and push them together into two groups of shapes: a group of non-native species and a group of native species. The resulting collection of shapes represents 100 percent of your plant biomass, and you should be able to see approximately how much of that comprises natives versus non-natives.

A couple of Omaha residents I spoke to said that their neighbors had complained to the city about their lawns, using city weed laws to challenge the presence of native prairie grasses in their yards. The following is the part of the city code used to challenge their native plants:

> It shall be unlawful for the owner, agent, occupant or other person in possession, charge or control of any lot or ground in the city, or any portion thereof, except for the Douglas County Land Reutilization Commission, the Omaha Municipal Land Bank and the City of Omaha, to do any of the following:

> 1) Permit, allow or maintain any growth of noxious weeds, grass or other worthless vegetation 12 inches or more in height.

> 2) Permit, allow or maintain any uncontrolled or uncultivated growth of brush, vines, woody volunteers, or other worthless vegetation which (i) prohibits the city from abating an existing offense; (ii) presents a serious health or safety risk to occupants, structures or the public; or (iii) offers vector or rodent harborage.

Is there anything in this code that sticks out to you as being intentionally or unintentionally hostile toward native plants in lawns? If so, what changes would you recommend to make city weed laws more accommodating to residents who want to incorporate native plants into their lawns?

That entire statement is hostile to the objective of living with nature, our only viable option going forward. It is also full of unsupported nonsense. Two that stick out are the statements about noxious weeds and native grasses as being "worthless vegetation." All states have an official list of noxious weeds. These are typically plants that have posed problems for agriculture in the past. Milkweed is usually on the list. So, for example, if we follow your town's ordinance, we can kiss the monarch goodbye. If we include other flowering plants not typically used in landscaping as worthless vegetation, we can also kiss our pollinators goodbye. "Worthless" vegetation? Worthless to whom or to what? That is totally subjective. In my view, turf grass is worthless vegetation, at least ecologically, and we have 44 million acres of it in this country. By this ordinance's definition, most of the plants of the world are "worthless." If we remove them, humans will be gone in a few months through total ecosystem collapse.

The second statement that bothers me is the claim that plantings other than lawn "present a serious health or safety risk to occupants." Show me the data that demonstrates that native plants in your yard pose a serious health risk. It can't be done, because *no* data supports this. And if they think native plants create a tick problem, I'd counter with this: too many deer create a tick problem, not native plants. Native plants also have nothing to do with mosquito issues. Mosquitoes do not breed in plants; they breed in water. So we can dispense with the claim of "vector" issues.

That leaves us with the suggestion that native landscapes create "rodent harborage." They do create habitat for field mice and voles, both essential components of most terrestrial food webs. But I suspect the ordinance writers are talking about house mice and rats, both introduced from Europe and closely associated with human garbage. Trust me; the bubonic plague was not the result of too many native plants and not enough lawns! The ordinance as it is written is full of inaccurate scare tactics. It needs to be totally rewritten in view of the biodiversity crisis that really does pose health risks to us all.

I am a municipal forester in Maryland, and a resident cites your work as their inspiration in advocating for only native trees to be planted along the street. Currently, about 70 percent of the roughly 1650 trees that we plant each year are native to the United States, but we still do plant some non-natives, mainly in unforgiving locations and upon request from adjacent homeowners. Roughly half of the larger growing species that I plant are oaks, and I do this at an expense to both our budget and our reputation. The white oak group consistently has one of the lowest rates of survivorship past the first year, and we are trying to get away from the red oak group due to high mortality from bacterial leaf scorch. Because of both funding restraints and lack of demand from residents, we currently replace less than one tree for every tree that is removed along the street. Do you advocate that municipalities plant only native street trees, and have you considered the implications of such a policy—mainly long-term survivability, lower diversity, increased future backlogs for residents to receive tree maintenance due to the elimination of non-native species with notoriously low-maintenance needs?

No, I don't insist that all street trees be native, but I certainly encourage municipalities to consider natives first. You may be planting 70 percent natives, but many municipalities don't come close to that. In Portland, Oregon, for example, more than 90 percent of the street trees are from outside of the Pacific Northwest. I notice that you say 70 percent are native to the United States, and that gives me pause. You could plant a street with 100-percent Douglas fir, Colorado blue spruce, and Kentucky coffee tree, all of which grow regionally somewhere in the United States but none of which are native to Maryland. *Native* is an ecological term, not one with political boundaries. A native tree has an evolutionary history with the ecosystem in which it is planted. Let's not forget why we are encouraging natives in the first place: to sustain the food web that enables us to share our human-dominated spaces with the "birds and the bees." Planting a species form the West Coast in Maryland will not help support the local food web because that species will not support the insects that drive that food web, even though the species may be from the continental United States.

Is there room for compromise? Absolutely. Our research has shown that food webs can tolerate up to 30 percent of the woody plant biomass in an area being non-native without collapsing. So an occasional ginkgo or dawn redwood will not sink the system, even if it doesn't add much ecologically. But for obvious reasons, I will never support planting invasive street trees like Callery pear, Norway maple, or Chinese elm, no matter what the local resident requests. That is just not good Earth stewardship.

Have I considered the implications of tree policies? Of course! Right plant, right place applies to natives as well as non-natives. I do not promote planting species with a low chance of surviving, with a shorter life span, and with higher maintenance requirements. But if you are asking me to believe that all native street trees are higher maintenance and have lower survivability than non-natives, you are asking a lot. The beloved Callery pears and Chinese elms, both famous for disintegrating during ice storms, have some of the highest maintenance requirements of any street trees. Many native street trees are long-lived and low maintenance when planted in appropriate places. How the tree is planted is also an important issue. I suspect you have poor survival rates with your white oaks because you are planting expensive, large trees that do not have a root mass large enough to support the aboveground biomass. In that case, the expense and survivability of the oak aren't the problem; it's the municipality's insistence on planting only large trees. I know the arguments for planting large trees: instant gratification, mower damage, and so on, but smaller trees are less expensive, and when they have a chance to lay down a normal root system, they grow quickly and are far healthier adult trees than large balled and burlapped trees or those grown in air pots. Surely this is an area where we can seek compromise.

Will favoring native trees lead to lower diversity? Maybe, if you plant only a few species of natives, but there is no need for that. I talk a lot about oaks because, ecologically, they are the best trees in 84 percent of the counties in which they occur naturally. But I am not promoting the exclusive use of oaks. We have oaks, but we also have maples, birches, redbuds, eastern red cedars, sweetgums, fringe trees, silverbells, native pines, sycamores, ironwoods, basswood, elms, and more, that can all be used as street trees, and they are all native to Maryland. Building diversity through non-natives sounds good but offers little ecological value, because most of those species do not contribute much to local ecosystems. Diversity is ecologically meaningful only when the species that create that diversity contribute ecologically.

Are there any non-natives that score higher in terms of supported lepidopteran species? I recall that ginkgoes score very low, and I now only plant them by request. Are there other non-native species that I should consider avoiding?

Nearly all non-native trees support very few Lepidoptera. You can add crape myrtle, Norway maple, dawn redwood, deodar cedar, kousa dogwood, styrax, and ornamental cherries like Kanzan cherry to the list of low performers. That said, a few non-natives do support some Lepidoptera. Non-native willows, such as weeping willow, do well, as do some ornamental crabapples. They are so closely related to native species that many North American moths and butterflies can't tell the difference. But there are no non-native species that support more Lepidoptera than native species.

Can you give us some tips about planting those challenging areas between the sidewalk and the street known as "hell strips"?

Planting so-called hell strips is not much different from planting any other confined space, with one major exception: soil quality. But before we get into that, step one should be to check local ordinances, which often restrict the size of the tree you may plant, particularly if overhead wires are present. Otherwise, you should be free to do what you want.

Hell strips can serve simultaneously as pollinator gardens and rain gardens.

Once you're cleared to plant, though, you should take a good look at the existing soil. Often it is just an extension of the type of soil you have in your front yard, but every once in awhile, a builder will use a hell strip to dump building waste, including chunks of concrete, sheetrock, nails, empty paint cans, and other such "soil additives." The builder then spreads a thin layer of soil over the junk to hide his ethical shortcomings. If you are an unlucky recipient of such a practice, there is no getting around it; you will have to dig it all out and replace it with real soil. From then on, it's just like planting any other space.

Assess the amount of sun and water the area gets, design the space (tallest plants in the center and then shorter plants on either side), and then plant. When bordered by sidewalk and street, hell strips can be either drier or wetter than surrounding soils, depending on the drainage slopes. But don't worry too much about those scary details. Try the plants you want. If they don't do well, try something else. It won't take you long to find a plant palette that works for you and that space. You can get much good advice on planting hell strips from Nancy Lawson in her book *The Humane Gardener* and from Evelyn Hadden's book *Hellstrip Gardening*.

What about restoration options for those living in the city who may not have a lawn but rather a patio or rooftop?

This is an important issue, because 82 percent of the US population now lives in cities. Fortunately, they too can help regenerate biodiversity in two ways: by creating a container garden and by creating a planting under an existing tree. Container gardening with flowering plants needed by specialist bees and migrating monarchs can turn a lifeless building into an important resource for these creatures. Many natives do well in patio pots, including Joe Pye weed, various milkweed species, and asters. City dwellers can also adopt a tree on the grounds of their apartment or condominium and, with permission from the superintendent or HOA, create and maintain a large bed under the tree. Any grass under the tree will be removed, and so will the need to mow under the tree, which will prevent soil compaction exactly where the caterpillars that develop on that tree need to wiggle below ground to pupate. Mulch under the tree can be a nice bed of leaves, and native groundcovers can be planted to

Even city-dwellers can plant for pollinators using containers with native plants on porches and balconies.

keep those leaves in place. The tree will appreciate the efforts, and if everyone in the complex adopted a tree, it would reduce the amount of lawn on the premises as well.

I am increasingly worried that the trees in my yard will blow down and hit my house or my car. Should I cut them down?

I don't blame you for being worried. After every storm, the media focuses on trees that have fallen on houses or cars. The impression we get is that all trees will fall soon and crush your property. If the news spent equal time showing you the thousands of trees that weathered the storm without falling, you wouldn't be as worried. But occasionally a tree is blown down, particularly after a lot of rain—so when should we worry? The risk posed by your trees will depend on what kind of trees you have, whether they grew straight and tall while in a forest but then became isolated when your house was built (very common), whether they have grown in a wet area (trees that grow in wet soil have much smaller root systems), whether they are growing near other trees so that their roots are interlocked with those trees to form a stable root matrix, and whether the trees are diseased and dying.

You may be surprised at the results of recent studies that show that healthy trees actually protect homes from wind damage during hurricanes. A study in North Carolina carries the headline, "Don't Cut Down that Tree: Preserving a Protective Buffer Against High Speed Winds." Researchers at the University of Florida agree: "A healthy urban forest with a mixture of young and mature trees provides benefits such as protection from high winds." A specialist in North Carolina warned that "If your neighbor clears all his trees, the added open space can put more stress on your roof, and on your trees."

We are about to begin removing invasive trees to make room for the arboretum we are building on our property. We also need to thin out some trees to give the trees in the collection some room. Many are within 5 feet or closer. I remember you saying that trees intertwine their root systems, making them stronger and more resilient. Is there a rule of thumb for how close or far away is best? We have two giant oaks that are within 6 feet of each other. Is it best to leave them be, or take one down?

There's no hard-and-fast rule of thumb. It is easy to find large trees growing right next to each other in a forest, but I think any trees within 10 feet of each other have intertwined their roots enough to give them added strength during a windstorm. Please leave both of your giant oaks. Just the fact that they are giants shows they are doing well where they are.

Are there benefits to small plantings, such as in very small yards or areas around public buildings, schools, and so on? If so what are those benefits?

Small plantings help mobile insects, particularly pollinators and butterflies, and can serve as vital stopover sites for migrating monarchs. Don't view your small planting in isolation. Your small planting will add to your neighbor's small planting, and their planting will add to their neighbor's planting. In combination, they are not isolated, small plantings anymore, but a growing network of plantings that will make an important difference for many creatures.

I live in Washington, D.C., and I have a problem with rats at my bird feeder. Any suggestions?

Many of our cities have large Norwegian rat populations because we leave so much of what they eat lying around. That's garbage, for the most part, but rats are omnivores and they love bird seed too. What to do? Unfortunately, there isn't much you can do except periodically stop feeding the birds. The rats will soon realize that there is no longer a steady food supply in your yard, and with a little luck, they'll move on. Trapping the rats can slow them down, but it's a pain in the neck (or their necks) and is not a permanent solution because other rats from the giant city population will move in before long. One thing you should *never* do is use rat poison. It does kill rats but not quickly, and after the rats are full of poison, they wander around aimlessly outside and become easy prey for hawks, owls, foxes, and bobcats (yes, even in cities), all of which eat poisoned rats and then die gruesome deaths. These animals are the very predators we need to keep our rabbit and vole populations in check.

You promote oaks, yet they tend to be huge trees at middle and old age and are thus less suitable for an increasing urban/suburban population. What species of trees are perhaps more sustainable for a densely populated, warming globe?

Depending on where you live, there are many top-performing native options besides oaks. In order of their value to wildlife, these include native species of willows, cherries and plums, birches, aspens, maples, alders, hickories, elms, pines, basswood, and hawthorns.

What is the minimum effective leaf-litter-island garden under a tree? I'm trying to reach a compromise amount of "unraked mess" and maximum insect habitat.

This question is the target of graduate student Emma Jonas's research at the University of Delaware. Leaf litter provides several ecological benefits that raked lawn does not: it protects the community of soil organisms whose job is to recycle nutrients the tree that produced it used the previous year, and it enables, through fungal associations, the tree's roots to use the nutrients again in the future. Leaf litter also protects the soil itself from becoming sun-baked, compacted, and eroded by wind and water. And, as you say, leaf litter is habitat for insect predators such as fireflies and ground beetles, spiders, pest-controlling parasitoids, and caterpillar pupae.

Emma's experiments suggest that the limiting factor for caterpillars that pupate in the soil is soil compaction, so from that perspective alone, creating no-go zones—areas where we neither walk nor mow—under our trees will improve caterpillar survival. It is clear that the bigger the beds under your trees, the more of these ecological benefits will be provided by retaining leaf litter. That said, I would think pushing your beds out to the dripline of your trees' canopies would be a good start. Do those beds have to be unraked messes? Of course not! The ideal bed is mulched with leaves overplanted with one or more attractive groundcovers. There are many to choose from, including foamflowers, wood poppies, native pachysandra, wild blue phlox, goldenseal, golden ragwort, mayapples, Virginia creeper, ferns, and others.

Lawns

How does one successfully recruit neighbors to join in the efforts to create more native habitats in their yards?

This is a challenge I have thought about for the last 20 years. I have not yet discovered any single approach that works all the time, but I can pretty much guarantee that proselytizing won't work. Imagine how receptive you would be if your neighbors told you that you weren't living right and that you should be more like them. But there *are* ways you can send a message, and they do occasionally work. The best is to be a good example by building a landscape that is simultaneously beautiful *and* ecologically functional. What your neighbor fears most is lowering their property value. If you can demonstrate that this won't happen, they might alter their biases. Several studies have shown that adding trees to your landscape actually *increases* property values by from 3 to 15 percent! Money talks!

You are fighting a long-standing cultural norm that has erected big lawns as one of our most important status symbols. What counts most, though, is not the size of your lawn, but how well it is manicured. There is still the perception that native landscaping is the same as *no* landscaping. It is anything but. Regular mowing convinces your neighbor that, even though you have less lawn, all is well in the neighborhood. Another approach that is surprisingly well-received is a sign designating parts of your property has habitat. National Wildlife Federation, Homegrown National Park, Audubon, and other groups all can provide signs to post in your yard to explain that your native gardening style is intentional and that it will help the ecosystem your neighbors depend on—as well as our beleaguered pollinators, birds, butterflies, and more.

You talk about reducing the area now in lawn. How exactly does one remove turf grass?

Many books have been written about how to change lawn into meadow, and they all agree on one thing: step one is to kill the turf grass (right down to the roots). If you don't do this, you will face weeding issues forever, and if you are starting from seed, your turf grass will easily out-compete your seedlings and you will have wasted lots of time and money. So how do you kill grass roots? Through shading, heating, digging, or applying an herbicide. If you stretch black plastic over the grass in question, you will block the sun your grass needs, and under the right conditions, it will get hot under that plastic and kill the grass. You should leave the plastic in place for several months. Similarly, spreading builder's paper over the grass smothers it and has the added advantage of not needing to be removed, unlike plastic. Builder's paper eventually degrades and you can plant plugs right through it.

Cardboard mulching is yet another approach to smothering your grass. This is just what it sounds like: you smother your grass with layers of cardboard and compost that break down after five months or so and become part of the humus layer of your new bed. The top layer should be a thick layer of compost to keep the cardboard from drying out. If the area you are treating is large, you will need a lot of compost—more than most people have on hand. But a thick layer of leaf litter will do the trick. Pro tip: Be sure to remove all of the packing tape that is usually on cardboard boxes. It does not break down, so if you don't first remove it, you will be pulling old tape out of your bed for years. Of course, you could just dig down far enough to remove the sod and all of its roots. This is fast but labor intensive, and you also lose a lot of topsoil that is tied up in the roots of the sod. You can also rent or buy what is essentially a flame thrower. If you like to play with lethal gadgets, this method of cooking your lawn to death is for you.

Finally, you can do what the professionals who do not have the luxury of time do: spray the target area with an herbicide specifically formulated for killing grass. This also works quickly but has to be done correctly and responsibly. The misuse and overuse of herbicides has given them a bad name, but when they are used sparingly and correctly, they are a valuable restoration tool. If you choose this route, first let your grass grow to about 6 inches tall. Select a product based on glyphosate, but do not choose one labeled "extended control" or "season-long." Those products keep killing far longer than necessary. Also, do not spray when rain is expected within 48 hours. You don't want any herbicide washing into the nearest stream or river. Follow the directions on the label, and never use more product than is needed.

Once your grass is dead, you have two choices: seed with a meadow/prairie mix appropriate to your area or plant plugs. If you are creating a large meadow, plugs will be prohibitively expensive, but they are perfect for pocket prairies and other smaller plantings. Do your best to buy seed (or any plant) from close to your provenance (the longitude, latitude, altitude, and environmental conditions in your area). Many seed suppliers maintain seed stock from different areas around the country. Books by authors Heather McCargo, Owen Wormser, and Neil Diboll and Hilary Cox all provide excellent guidelines for how to proceed next.

In addition to all of those options, if you are a patient person, you might consider reducing the area of lawn you have over time by planting a tree and building a bed around the base of the tree. As the tree grows, you can expand the bed at least to the dripline. This reduces the amount of work needed to remove lawn because it spreads the effort out over time, and it's less of a visual shock to your neighbors. As I write, I am looking out my window at an oak tree I planted from an acorn 22 years ago. It now has a canopy spread of 40 feet and covers a bed beneath it with an area of 1247 square feet. In effect, adding that one tree to my yard has reduced my lawn area by 1247 square feet, with no labor on my part!

My husband loves mowing the lawn and I love raking leaves. What ecologically valuable things can we replace these activities with?

There is nothing wrong with raking leaves off parts of your lawn into areas where you no longer want lawn. And the lawn you keep still needs to mowed and cared for. But dedicating acres of property to lawn just for the entertaining value of mowing is something we must move beyond. By keeping some of the lawn, you send a message to your neighbors that you understand the neighborhood's landscaping culture and you are not trying to buck it. But you're going to have less lawn and more plants. Any chance you and your husband can get interested in planting trees or creating new beds? Or maybe watching and photographing the wildlife that visits your property?

Would you recommend, if money is tight, that we allow a section of yard to grow without planting natives? Wouldn't that be better than mowing the lawn? This also would satisfy the challenge of whether to water. Is there a guide to local natives that use less water? (I am in southern Oregon, currently in heat wave.)

Not mowing would be better than mowing, simply because an unmowed lawn will absorb more water when it does rain than a mowed lawn. Also, if you don't mow for long enough, some native forbs will start to come in. But neighbors often complain about unmowed lawns, and mostly what you'll have is a tall, non-native grass species that won't support much wildlife. In my opinion, it would be better to reduce the area you have in lawn and plant the areas you take out of lawn in appropriate drought-tolerant natives. If you don't get enough rain to support a lawn without supplemental watering, then forget the lawn altogether and move entirely to xeric (drought-tolerant) plantings. You can formalize the area with gravel paths. Adding natives to your landscape does not have to be expensive. You can reduce costs by buying small plants (they will grow!), starting from seed whenever you can, and adding plants slowly over time. Even adding one or two plants a year will make a big difference before you know it.

Our church group created several native plant gardens. We want to convert even more grass to native gardens, but now I am hearing conflicting information about the efficacy of traditional lawns for sequestering carbon. Is carbon lost when grass is cut? Does the height of grass make a difference?

The lawn industry is starting to feel threatened by the national movement to reduce the millions of acres now in turf grass, and it has begun to tout the ecological value of lawns. But that is a debate they cannot win if they stick to the facts. When it comes to accomplishing the four ecological responsibilities of every landscape (supporting pollinators and the local food web, managing the watershed, and sequestering carbon), turf grass is the *worst* plant choice imaginable. Let's just focus on carbon capture. Yes, turf grasses, like all plants, build their tissues out of atmospheric carbon. But how long does grass store that carbon? In the average yard, about a week. As soon as we cut the grass, we release all the carbon that was stored in grass leaves since the last mowing. If we let the grass get longer, by reducing mowing to once every two or three weeks, mowing still releases sequestered carbon as soon as the cut grass blades decay. We need to capture carbon and store it for centuries, not weeks. And don't forget that if you use a gas-powered mower, which most of us do, mowing adds carbon dioxide and other pollutants to the air. Finally, although grass roots do add carbon to the soil over time, any plant species with a bigger root system does this much better. Grass has very short roots (dig some up and see for yourself), which is why you have to water your lawn during dry periods, wasting precious water and energy that could be better used elsewhere. Bottom line is this: I would continue to encourage your church group to reduce the amount of lawn wherever people don't walk frequently.

My question is around the issue of two million acres in the United States being devoted to golf courses. I'll confess to being a pretty avid golfer and formerly thought of these public lands as not being "developed." Do you think golf courses are just dead zones in terms of biodiversity, or would it be impactful enough to plant local native plants throughout, between holes, along waterfront, and so on, while keeping the primary playing area grass—just the fairways and greens?

Golf courses could do a lot to support biodiversity. And you outlined most of it. Keep the fairways in manicured grass, but push the canopy trees, the rough, the ornamentals near the clubhouse and parking lots, and so on, toward natives. Most golf courses already have lots of non-native plants; if we change them to productive natives, few club members would notice, but the local biodiversity would notice quite a bit. Of course, reducing unnecessary use of insecticides would help too, as would changing night lighting to bulbs near the yellow spectrum to reduce needless insect kills.

What does an alternative natural lawn look like, and how do I create one?

To me, the concept of a natural lawn is somewhat oxymoronic for the simple reason that lawns aren't natural. My feeling is that the main purpose of a lawn is to provide an area for us to walk through our landscapes without killing what we are walking on. Turf grasses are excellent at this and can take even fairly heavy foot traffic without dying. To maintain lawns, even if they are just paths or well-defined avenues, we have to mow; otherwise, the grass quickly becomes so tall that walking through it is difficult. And no matter what plant species we use in a lawn, mowing will kill whatever is trying to live in and on those plants. If you have decided to replace your lawn with some other type of planting, that's great, and I congratulate you—but then that's really not a lawn any longer.

Krissy Boys of Cornell Botanic Gardens experimented with some native grasses in 2009 to create lawnlike areas that did not require constant mowing (although they did need to be mowed a few times a season). She found two species of *Danthonia* oat grass that worked well as alternatives to traditional turf grasses. To these low-growing grasses, she added several species of short perennials (including creeping phlox, pussytoes, heartleaf foamflower, wild geranium, various short sedges, blue-eyed grass, and bluets) that thrived in sunny, dry conditions to enhance the wildlife value of her creation. None of these plants can take even moderate foot traffic, but if you are trying to create a lawn alternative that requires little maintenance, remains short for weeks at a time, and provides varied resources for pollinators, they work well.

One alternative to the current standard (a weed-free expanse of grass that requires regular herbicide and fertilizer treatments to meet neighborhood expectations) is to add white clover to your lawn. Any readers approaching my age will remember that most lawns of the 1950s included clover. White clover adds nitrogen to the soil, so fertilizers are no longer necessary. It also blooms all summer long, much to the delight of generalist pollinators such as honey bees and bumble bees, even with regular foot traffic and mowing. It is also cheap and easily established. Most hardware stores sell clover seed; buy a bag or two, and in spring broadcast that seed over your lawn (grab handfuls of seeds and throw them), and that's all there is to it. Adding white clover to your lawn is the easiest way to enhance the wildlife value of your lawn, and you eliminate the need for fertilizers and herbicides. The lawn industry does not promote clover because it wants you to buy lots of fertilizer, which is laced with herbicides. But your watershed would vote yes to white clover if it could.

When converting to natives, should I just stop mowing? Won't the neighbors hate me?

Stopping mowing will not give you a native landscape. It will just give you tall non-native turf grass, and it very likely will displease your neighbors. I am proposing that you reduce the area you have in lawn by planting trees and shrubs and putting beds with dense groundcovers under those plantings. The lawn that remains after you add these plants should be manicured as always (minus the unnecessary addition of fertilizer and herbicides). It will serve the

A densely planted landscape can still meet today's cultural norms if it is lined with a manicured strip of turf grass.

very important social function of being what Sue Barton of the University of Delaware calls a "cue for care." How you treat your remaining lawn will demonstrate to your neighbors that you understand the cultural expectations of your neighborhood's landscapes. You will dutifully care for your lawn; you will just have less of it.

Coming from California, where we don't get a lot of rain, I'm wondering if replacing a lawn with native plants uses less water.

Absolutely! Nobody was watering the plants in California before humans came along, and the native species did just fine. Ask the California Native Plant Society for a list of species they recommend for where you live. And the State of California offers a three-dollar rebate for every square foot of lawn you remove and replace with xeric plants.

Leaf Litter

What is better: fallen leaves, bark mulch, or wood chips?

Nothing beats fallen leaves. They are the right weight, they fall in the right amounts, and they contain the nutrients your trees used that year and will recycle them into the soil so that your trees can use them again in the future. Leaves also maintain the soil moisture and humidity perfectly for the hundreds of species of detritivores and mycorrhizae that live in the soil and enable your plants to thrive. Bark mulch and wood chips are better than bare soil, but not nearly as good for the soil as leaf litter.

I've heard you state that it is a "loss" to remove all your fallen oak leaves in autumn. Each year, my yard receives and I gather much more fallen leaves than my property can absorb. I mulch my beds deeply, but many cubic yards of leaves are still available to the city for curbside disposal. Sometimes I shred piles to reduce the volume by running my lawnmower over them. What sort of regimen can I devise for saving them until spring, while keeping them whole?

Every time I hear this question, it reminds me of a phone call I got from my son. He had recently bought a new house and was dealing with autumn leaves for the first time.

> Him: Dad, I have too many leaves. What should I do with them?
> Me: Rake them into your flowerbeds.
> Him: I don't have enough flowerbeds.
> Me: Exactly!

I encourage people to keep the leaves that fall on their properties, on their properties for four reasons. They are homes for many creatures, especially moths that have spun their cocoons among them. Most of these moths are tiny but are still important components of the food web. But some, like the luna moth, are among our largest and most beautiful insects. They do not enjoy being run through a lawn mower. Another reason is that fallen leaves are a food source for the caterpillars of 70 species of moths commonly known as litter moths, as well as for several beautiful lycaenid butterflies. It's hard to imagine caterpillars living off of dead leaves, but they do. So when we dispose of our leaves, we destroy the Lepidoptera community in the leaf litter. A third reason—and from the perspective of your trees, the most important one—is that many of the nutrients your tree used the previous growing season are in those fallen leaves. In the forest, leaves are broken down by hundreds of species of animals, bacteria, and fungi, collectively called detritivores, which return the nutrients stored in the leaves to the soil so that they can be used again by the trees. When we remove leaves from our properties, we are slowly starving our trees. Finally, leaf litter creates the moist conditions that protect the soil community. More species live in the soil than above the soil, and all of them require high humidity to survive. When we remove the blanket of leaves from the soil surface, the soil dries out, heats up, and/or is eroded away by wind and storm water.

Many moths, including the luna moth, spend fall, winter, and spring within a cocoon nestled in leaf litter.

So how can you keep intact leaves on your property? The best way is to expand the beds under your trees so that they can absorb more of the leaf load. Expanded beds give you the added benefit of reducing the area you have in lawn. Most people have far more space dedicated to lawn than to beds under their trees. We need to reverse that; more area should be in beds than in lawn. Rake the leaves around your trees, where they can form a protective blanket for all of the tiny organisms that reside in your soil community all winter long—and all spring and summer too. Make the beds as big as possible; extending them at least to the trees' dripline is best. Rake the leaves into those beds in fall and then plant right through the leaves in spring.

How do I balance maintaining a duff layer of decaying material for natives to hibernate in with the need to keep down hibernation areas for garden pests?

It is my humble opinion that reducing duff to reduce pest populations is somewhat of an urban legend. Somehow, those pests always find our plants, even when we have no duff (a very unhealthy condition for our plants and soil community). What really controls pests are natural enemies, so designing landscapes that support all trophic levels of enemies (predators and multiple levels of parasitoids) is the way to go. Besides loss of habitat, the number one destroyer of natural enemy communities is insecticides. Every time we spray, we kill far more of the natural enemies and insects that are not pests than the pests themselves, which then enables pest populations to explode. Bottom line: leave the duff!

If you don't rake leaves away from your beds, will voles eat all of your perennials?

Who was raking leaves away from the thousands of species of native perennials in North America before we came along? Obviously, voles did not eliminate perennials back then. And what will happen to the soil community your perennials depend on if you rake all of your leaves away? Most people remove leaves but then replace them with expensive mulch, in hopes that the mulch will perform the same ecological duties as the leaves. It won't. The point is, if you create a landscape so inhospitable that voles cannot live there, you have a dead landscape. Voles are natural parts of meadow ecosystems. They will eat a perennial now and then, but lots of things eat voles as well, including foxes, owls, hawks, and black rat snakes, all of which should be tolerated so they can keep your vole population in check.

When I rake my leaves, they end up on my compost pile. Does this harm the pupating insects?

Whether or not putting leaves in your compost pile harms pupae within them will depend on how deeply they are buried in the pile. Pupae near the surface should be fine. I don't want to discourage you from composting, but the leaves that remain in your planting beds will be the real contributors to your property, as they provide a protective blanket and food for the soil community beneath your trees.

Does cedar mulch repel butterflies and other beneficial insects, or is that a myth? All this time I've been using cedar, and now I'm concerned.

I'm pretty sure that is mostly a myth. Moths and butterflies find their host plants by their smell. If cedar chip smell is so strong that it masks the volatiles from host plants, then the chips would deter moths and butterflies. But it's my experience that cedar chips quickly lose their odor when exposed to sun and rain.

What's the best way to manage roadsides to support pollinators?

The best way to manage for pollinators is to replace the omnipresent grass growing along our roads with the flowering plants that pollinators need. Road managers want at least one wide mower's worth of grass to enable cars to pull off if necessary, but the rest of the verge is best managed in some type of meadow/prairie planting. Townships on a budget can push their grass verges toward native bunch grasses and forbs such as goldenrod, perennial sunflowers, and asters, just by shifting their mowing schedule. Nothing needs to be planted. They can mow two or three times during spring, but then stop mowing completely until the next spring. After a few years of that mowing schedule, the cool-season European grasses we use for turf will start to be replaced by warm-season natives. Whenever reduced mowing is suggested, there is pushback from the mowers, who fear they will lose their jobs. Over time, the mowing staff will be reduced, but guarantees should be made that no one will lose their job. Cuts can be made simply through attrition. And, by the way, studies have shown that the oft-cited need to mow for safety reasons is unfounded. There are not more encounters with animals when verges are left unmowed.

I'm studying sustainability management and policy at Penn State. I'm interested in your thoughts on driving systemic change to regenerate ecosystems. Homegrown National Park is encouraging individual collective action. My question for you is, how do we push for systemic regenerative change?

Several top-down inputs could help with this cultural change. Some are already happening. I'm referring to the ban on selling invasive plants. Some states have taken up the legislative "stick approach" (punishing those who sell invasive plants), which is always way too late to make a difference, but this is helpful in the cultural change that is so necessary. But the "carrot approach" is sure to work better: incentivize people to change the way they landscape. Water shortages in the West have stimulated state-supported rebates for converting lawn to sustainable landscapes, and they have been well-received. Minnesota and Pennsylvania cost-share programs are focused on reducing watershed pollution from lawn care more than on saving water. Even public utilities in a few states are giving people 100-dollar coupons to replace thirsty non-native plants with water-efficient natives. The USDA Conservation Reserve Program (CRP) incentivizes farmers to plant prairie strips right through their corn and soybean fields. Such plantings not only provide valuable resources for pollinators, but they also reduce topsoil loss by 95 percent and water pollution by 90 percent. And the cities of Saint Louis and Fayetteville, plus the state of North Carolina, have put a bounty on the highly invasive ornamental Callery (Bradford) pear. If you take out a Callery pear on your property, you get a free tree replacement.

But much more could be done. Throughout much of the country, mega-farms have removed hedgerows that could provide safe harbor for biodiversity in an otherwise dead landscape, and roundup-ready crops have enabled farmers to kill all of the roadside plants bordering their fields and replace them with turf grass. This culture shift alone has brought the monarch and many of our native bees to their tiny knees. But we could incentivize farmers to reestablish hedgerows wherever possible and to put the "weeds" (productive native plants such as asters, goldenrod, evening primrose, and milkweeds) back along roads bordering agricultural fields.

We could initiate national tax incentives for reducing lawn and using keystone native plants that support most of our nation's animal life. That alone would do much to stem the loss of species from human-dominated landscapes. Tax incentives would send the message that ecological landscaping is important and supports the greater good for everyone. It would also discourage HOA rules that discourage homeowners from responsible landscaping, and it would give people cultural permission to do the right thing without being a "troublemaking rebel." What could pay for these programs? I have long favored a pay-per-use fee, required of every American. Just as we don't expect to get services from an online streaming service for free, we should be willing to pay when we use Earth's resources. Our natural life support system is not free; it requires ecological capital to produce and it is finite. We all use ecosystem services every day, so what if we made the earth-service fee 10 dollars per year per person? If we had done that as of January 1, 2022, we would have generated 3.32 billion dollars a year in the United States alone that could be used to support ecologically sound programs. That's 10 dollars per year! That's two medium-sized double-chocolate mochas! Isn't the ecological health of our planet worth more than a couple cups of coffee?

Our property on Cape Cod is mainly natural woods, thick with oaks and pitch pines. We want to keep it that way. However, piles of oak leaves under each tree can be unsightly, especially because a number of trees we have are in groves and near the areas of foot traffic and the areas where we like to lounge in summer or grow our vegetables. What techniques would you suggest for cleaning the areas with least disturbance to the wildlife? What do we do with all those leaves? What composting "techniques" would be necessary to maintain natural balance?

Sounds like you have a great property. You can compost leaves in a single pile and use the compost for gardening purposes, or you can spread the leaves thinly throughout the woodsy part of your yard. That way, whatever organisms that are overwintering in those leaves can survive. Another option, if you don't want to see the leaves under the trees, is to plant groundcover consisting of low shrubs and perennials. Shrubs such as beach plum, bearberry, common juniper (*Juniperus communis* var. *depressa*), highbush and lowbush blueberry, inkberry (especially in wet areas), New Jersey tea, and Virginia rose are all good options. Many perennial groundcovers will completely hide those valuable leaves, including mayapple, foamflower, wild ginger, many species of ferns, violets, Virginia creeper, and native pachysandra. And, once established, all of those plants would enjoy a solid layer of oak leaves at their feet.

A diverse groundcover planting will hide the leaves that fall from your trees each autumn, while preserving their valuable contributions to the soil community.

CHAPTER NINE

Supporting Wildlife at Home

Birds

Is it good to feed the birds?

When done properly, feeding birds certainly is good—at least during winter and spring. Our bird feeders attract species that spend their winters in our yards (migrants that fly south for winter are insectivores and therefore not interested in seed). Most non-migrants depend heavily or entirely on seed and the fats they get from suet cakes, both of which are in very short supply in a typical residential landscape. Many common attendees at our feeders such as chickadees and titmice spend half of their time hunting insects and spiders, even in winter, but a well-stocked feeder can mean the difference between whether they make it through winter on your property or not.

I have heard ornithologists claim that you don't have to feed the birds because there is plenty of food out there. That statement baffles me, because typical landscapes are dominated by lawn and have no meadow plants loaded with seeds. Even if you plant your yard with appropriate seed-bearing natives, chances are your neighbor does not, and what you provide is not, in itself, enough to support entire bird populations. Remember that before we humans came and took over, 100 percent of the land was productive and there really *was* enough food to support huge bird populations. But those days are long gone. Studies have shown that birds with access to feeders enter the breeding season heavier and can produce more eggs in spring than birds that haven't been fed from feeders. The one downside to feeding birds is the potential to spread disease. This potential increases if you crowd several feeders into a small space and do not keep them clean. But a good cleaning once or twice during the season is enough to keep the birds happy and healthy.

Is spicy suet harmful to birds?

Spicy suet is one of many attempts to keep squirrels from eating bird food. Several years ago, someone discovered that birds cannot detect the spicy taste created by capsaicin, the "hot" compound produced by peppers. This discovery led to mixing capsaicin powder with bird seed in hopes that it would not harm birds but would be distasteful to squirrels. That approach worked but was messy, plus it was hard to mix the powder with the bird seed, and the pepper often got into the mixer's eyes, which was not at all a pleasant experience.

The next anti-squirrel invention was "spicy" suet. No mixing, no muss, no fuss, and a godsend for homeowners with serious squirrel issues. Does it hurt birds? There is no evidence that it does. I have watched several species of birds dig into hot suet for hours on end, and I know bird lovers who have used it for years with no apparent ill effects on their bird populations. If you are concerned about using spicy bird food, though, I can recommend a squirrel guard that really does work: the North States two-way squirrel baffle (large). I have tried most squirrel guards, and this is the only one that actually keeps squirrels off my pole feeder, as long as I locate the feeder far enough away from tree branches that squirrels could use to jump to the feeder.

Do birds specialize on particular caterpillar species the way insects specialize on particular plants?

Insectivorous birds are predators, and like predators all over the world, they cannot afford to be too picky about what they prey upon. Prey is usually difficult to find and capture, and there never seems to be enough caterpillars. So predators typically take whatever they can get. That said, birds do favor some caterpillars over others. Their favorites are the inchworms (family Geometridae) that used to be so common in spring. Inchworms are tasty and are not hairy or spiny. Birds avoid caterpillars that advertise bad taste with aposematic coloration (warning coloration that informs predators that the caterpillar is poisonous or otherwise dangerous) and those of species like monarchs that acquire the nasty compounds stored in their host plants. They also ignore hairy caterpillars unless they are really hungry, because the hairs of these species dislodge from the caterpillars and get stuck in the bird's esophagus and stomach. A hungry bird will take a hairy caterpillar, or one with lots of spines such as a buck moth, but before they eat it or feed it to their young, they spend several minutes whacking it against a branch to remove as many spines as possible. This necessary behavior considerably elevates the cost of hunting. There is no point in the bird preying upon something if it has to spend more energy preparing it for consumption than it receives by eating it.

Two bird species, Yellow-billed and Black-billed cuckoos, actually specialize on hairy caterpillars, including tent caterpillars, fall webworms, and spongy moth caterpillars. Cuckoos have evolved the remarkable ability to shed the lining of their esophagus and stomach periodically, ridding their digestive tracts of all the hairs that have accumulated there. Why did natural selection favor this trait? Because there are a lot of hairy caterpillar species out there, and many, like tent caterpillars and fall webworms, are gregarious. So if a bird finds one, it will have found hundreds! A bird that has evolved to eat hairy caterpillars will essentially have an unlimited food supply, a distinct advantage in the world of predators.

Can I assume that my hummers will feed only from appropriate native plants? Should I get rid of the non-natives?

Hummingbirds will use any flower (or plastic feeder) with a tube-like structure that delivers nectar and is easy for them to reach with their long tongues. Their relationship with plants is quite generalized. You do not need to get rid of non-natives that attract hummingbirds. But remember that 80–90 percent of a hummingbird's diet is insects and spiders, so you also need the plants that support insects and the spiders that eat them if you want to have nesting hummingbirds.

How should we treat (filter and sanitize) recirculating fountain water to reduce or eliminate diseases spreading among songbirds that use it? We

want to provide water sources for songbirds and we've found that bubbler fountains are particularly attractive to them, but we haven't found any guidance on water treatment protocols.

The key here is providing water that circulates. Water bubbling up through a rock and then splashing down, even a little way, oxygenates the water and helps keep it clean. Stagnant bird baths are a bad idea and birds rarely use them. They are attracted to water that makes noise, such as a babbling brook, because that signifies clean water. I don't think it's necessary to add anything to moving water to ensure cleanliness.

I put out dried mealworms for the birds in spring, and I notice that a couple of wrens seem to be using them regularly and flying the worms to their nests. I know there's nothing so good as a variety of fresh caterpillars, but are dried mealworms good for the birds?

Birds will hunt for food in the most economical way (that is, in a way that expends the least amount of energy). If you are supplying food, they will take advantage, even if what you're supplying is not the perfect food. I don't like dried mealworms because the innards turn to dust, leaving only the exoskeleton. Live mealworms are much better, but, as you say, caterpillars are best. Mealworms are the larvae of tenebrionid beetles, and they are very low in carotenoids, essential components of bird diets. That's where caterpillars out-perform all other types of bird food. They are very high in carotenoids and are perfect for nestlings that are growing so fast. The very best way to feed the birds is to load your yard with plants that make lots of caterpillars.

How can I make my yard raptor friendly?

The primary thing you need to have a raptor-friendly yard is raptor food in your yard—that is, good populations of white-footed mice and voles, rabbits, squirrels, and other birds, depending on the species of raptor you want to encourage. Sharp-shinned Hawks live off small birds like those that frequent feeders, so providing a source of clean bird food will help in that regard. Cooper's Hawks will take squirrels and larger birds such as pigeons, doves, grackles, and Blue Jays. Red-shouldered Hawks prey on snakes and frogs, so a small wetland will help feed them. Red-tailed Hawks go for larger mammals such as rabbits, but they'll often take rodents as well.

Generating good populations of these prey types often requires larger properties, but one thing everyone can do to help raptors is *never use rat poison. Never, ever!* And that goes for your pest control company as well. Rat poison does not kill quickly; instead, it weakens mammals that have ingested it so that they wander aimlessly outside and become easy pickings for hungry raptors, which soon become dead raptors. If you have a rat problem, old-fashioned snap traps work well.

How can I convince my neighbor to keep his cat indoors?

This may not be difficult if your neighbor is a good friend who is willing to listen to the facts about the ecological problems caused by outdoor cats. You can tell him that cats kill more than two billion birds in the United States every year—that's a third of all of the birds in North America—plus countless rabbits, mice, voles, chipmunks, snakes, toads, frogs, grasshoppers, praying mantids, and butterflies. Many people think predation by outdoor cats is "natural" and therefore nothing to worry about. But domestic housecats are not natural residents in North America, and there are a lot of them: an estimated 100 million cats spend some or all of their time outside pouncing on anything that moves. They impose a significant source of mortality for species that are not adapted to avoiding them. Cat lovers argue that it is cruel to keep a cat indoors. Toys and cat videos are useful forms of cat entertainment, and if you keep them indoors from birth, they will not miss outdoor living. For good tips on what to do if your neighbor is not open to your request, read Peter Marra and Chris Santella's 2016 book *Cat Wars*.

What are the best window treatments for preventing deadly window strikes by birds?

Every year, more than a billion birds fly into windows and break their necks. They are not committing suicide; they are flying into what they believe is open air, but which is actually a reflection in the window. What is a bit frustrating is that we know how to make window glass that reduces or eliminates these reflections. Yes, it's a bit more expensive, but many window-purchasers don't

This kestrel was permanently maimed by a housecat.

seem to value bird communities enough to pay the small additional amount to use it. What if part of our houses killed one billion (that is, 1000 million) dogs and cats each year? Do you think we would be willing to pay a few dollars more to end that carnage?

Non-reflective window glass should be code for all new construction. But what can we do about existing windows? Many groups sell window decals featuring hawks and other birds, in hopes that incoming birds will see them and veer off. I have tried these, and at least at my house, they did not noticeably reduce window strikes. But several other products do work. Ornithologist/ author Daniel Klem Jr. has devoted much of his long career to researching bird window collisions, and he reviews this work in his book *Solid Air*.

Caterpillars/Light Pollution

I hear you say that caterpillars are important. But why? Why are they more important than, say, beetles?

Though this book is about many things, the importance of caterpillars is the primary topic. I say caterpillars are important because so many other animals rely on them for food. Caterpillars, particularly those of moth species, are often called the "bread and butter of terrestrial food webs"—that is, they are an essential conduit through which carbohydrate energy, protein, and essential nutrients are moved from plants to animals. This is especially true in temperate zone ecosystems, where fruits are more seasonal and less rich in fats and proteins than they are in tropical ecosystems. Many animals, both vertebrates and invertebrates, do not eat plants directly; they eat consumers of plants—and not just any consumers. Caterpillars transfer more energy from plants to higher trophic levels than any other herbivore guild. Among vertebrates, caterpillar abundance, as well as the abundance of adult moths, helps maintain amphibian, mammalian, reptilian, and bat populations. In the United States, 42 of our 45 bat species are insectivores, and the majority of their prey comprises moths. And the number of moths consumed by bats is not trivial. Research by Elizabeth Beilke and Joy O'Keefe (2023) showed that when bats were excluded from oaks and hickories in Indiana, caterpillar populations increased three-fold. A 1998 study by Don White Jr. and colleagues found that a foraging grizzly bear can consume 40,000 army cutworm adults every day in early fall, and those moths contribute substantially to the bear's overwintering fat reserves. Black bears take advantage of caterpillars as well, eating up to 25,000 tent caterpillars per day. Dissections of prairie dog stomachs have revealed cutworm caterpillars in 14 of 15 individuals—and not just a few: a single prairie dog had eaten more than 600 cutworms, suggesting that caterpillars are a far more important food source for prairie dogs and probably other ground squirrels than previously thought.

Regardless of caterpillars' role in supporting other vertebrates, the best evidence of the nutritional importance of caterpillars comes from studies of birds. In the United States, most terrestrial birds rely on large numbers of insects during the breeding season. About 96 percent of North American terrestrial birds rear their young on insects, and with exceptions of entire bird families such as nightjars that specialize on adult moths, the protein- and fat-laden caterpillars are the largest component of nestling diets in hundreds of species of migrants and resident breeders. In fact, my former student Ashley Kennedy found that in 16 of the 20 bird families about which data exists, caterpillars dominate nestling diets.

Not only are caterpillars important components of nestling diets, but they are also required in great numbers. It takes many thousands of these insects to bring a clutch of birds to independence. For example, ornithologist Richard Brewer recorded Carolina Chickadee parents bringing an astounding 6000 to 9000 caterpillars to their nestlings, depending on the number of chicks in the nest, and chickadee parents continued to feed caterpillars to fledglings for 21 days after fledging. Similar data has been recorded for Wilson's Warblers, Bobolinks, Downy and Hairy woodpeckers, and 10 species of European passerines.

As important as caterpillars are for supporting birds, bats, and other vertebrates, they also serve as hosts for a staggering number of insect predators and parasitoids. Within the family Hymenoptera (bees, wasps, and ants) alone, now shown to be the largest insect order by Andrew Forbes of the University of Iowa, global estimated richness of the parasitoids that live off of caterpillars tops 700,000 species. Nearly all vespid (5000 species) and sphecid wasps (10,133 species) prey heavily on caterpillars, as do thousands of species of predatory damsel bugs, assassin bugs, stink bugs, and plant bugs. Tachinid flies may number 10,000 species, most of which attack lepidopterans. Among predatory beetles, seven genera of carabid ground beetles eat caterpillars. In short, a substantial component of animal diversity depends on the abundance, diversity, and general well-being of caterpillar communities. This, quite simply, is why favoring plants that support caterpillars and avoiding plants that do not is a great idea.

Is it the case that most native insects are specialists in that they can eat only one or two specific native plants? Or can most insects eat only one or two native plant families, which would mean many more than two specific native plants? Can you clarify?

The most accurate way to describe host plant relationships is to use the word *lineage* instead of *species* or *family*. Monarchs, for example, can eat only plants in the milkweed lineage. That does include a few species outside of the milkweed subfamily Asclepiadaceae, but they are all in the same lineage. It may help to think about all the species they *cannot* eat instead of trying to list the species they *can* eat. Even our most generalized caterpillars can eat only 7 percent of all plants available to them.

Are caterpillars dangerous?

In North America, the vast majority of caterpillars are harmless. A few species, such as the introduced spongy moth, have hairs that can cause an allergic reaction if they're ground into your skin. A few others, particularly those in the families Limacodidae and Megalopygidae (the slug caterpillars and flannel moths, respectively) and the buck moth caterpillars, have urticating hairs, which are really spines, and each one has a tiny poison gland at its base. If you brush up against them they easily break off, releasing the poison into your tiny wound. It hurts. But are any of those US caterpillars actually dangerous? Not really. But caterpillars in the tropics are a different story. I was once scheduled to meet a guide in Costa Rica. He didn't show because earlier in the day, he'd brushed up against a caterpillar with urticating hairs and was in the hospital!

You say *Prunus* species host a lot of caterpillar species, but I can't find any on either of my *Prunus ilicifolia* trees. There were signs of chewed leaves, especially on the older tree, but no caterpillars. I checked my toyons (*Heteromeles arbutifolia*) and found nothing on them either. I have had these plants for years. Shouldn't the moths have found them by now? My neighborhood is a desert as far as native plants are concerned. Is it possible that the moths that feed on the cherry and toyon are locally extinct?

There could be some local extinctions, but it's more likely that caterpillars are feeding on your trees at night and hiding by day. This is particularly true in hot, dry areas. Check again with a flashlight well after dark. Let's hope that is the explanation. A more insidious possibility is that your neighbor or neighborhood has hired a mosquito fogger. That will put an end to all of your insects.

I find the National Wildlife Federation Native Plant Finder super useful! But I can't find two commonly used plants—fall obedient plant and fothergilla—anywhere, no matter how many zip codes across the Southeast I enter into the database. As a landscaper, I get many questions from homeowners about their benefits (clearly they both attract bees!), so I'd appreciate knowing their benefits to Lepidoptera, if any.

Both fothergilla and obedient plant support only one caterpillar species, so they don't make it to the productive plants list in terms of supporting caterpillars. Also, fothergilla is often planted far north of its actual native distribution. Only plants that occur naturally within a county are included in the Native Plant Finder. Both are good pollinator plants though. Remember that the Native Plant Finder is designed for caterpillar production, not for pollinators. If you are interested in the best plants for specialist pollinators, visit National Wildlife Federation's Keystone Plants by Ecoregion website.

How do the larvae of the wavy lined emperor moth make themselves look like liatris flowers?

You are talking about the camouflaged looper caterpillar, which eats lots of types of flower petals and "sews" the petals onto its back with silk to hide from birds—a fascinating example of crypsis.

This camouflaged looper caterpillar has disguised itself using petal snippets from the flower it's eating.

How can I identify the caterpillars I find?

Not long ago, this was quite a challenge, but today you can use several sources that will get you close to—if not right on the money—identifying caterpillars in Eastern North America to species. Step one, though, is to do your best to take a good picture of your mystery creature. You will not remember its many physical traits long enough to match it to reference pictures if you don't have an image to refer to. David L. Wagner's *Caterpillars of Eastern North America* is a pictorial guide to 700 commonly encountered species across all families of the Macrolepidoptera group. The caterpillars of Microlepidoptera (leafminers, tiny fruit eaters, leafrollers, leaf tiers, and so on) are simply too tiny and similar to each other to be able to identify them with the naked eye. With help from others, Wagner also produced a book dedicated to the largest group of caterpillars, the owlets (*Owlet Caterpillars of Eastern North America*). The book includes great pictures that make it relatively easy to identify more than 800 species within the family Noctuidae. He has also been working on a guide to the caterpillars of western North America. There are many more caterpillar species in the West than in the East, so, as of this writing, he labors on.

There are digital options for caterpillar identification as well: iNaturalist and BugGuide are very popular citizen science–based apps that enable you to post an image of a caterpillar from anywhere so that amateur naturalists across the country can identify it for you. An interest in caterpillars often leads to an interest in the adult moths and butterflies they turn into. Good sources for identifying adult Lepidoptera include iNaturalist, *Field Guide to Moths of Northeastern North America* by David Beadle and Seabrooke Leckie, and *Field Guide to Moths of Southeastern North America* by the same authors.

Why are caterpillars such fussy eaters?

When I was five or six years old, I found a caterpillar crawling on the ground and asked my mother what I should do with it. She told me to put it in a jar and give it some leaves to eat. So I did—some grass and a few of the first leaves I could find. I'm sure there was no intentional lesson hidden in my mother's instructions, but one was learned anyway. Caterpillars do eat leaves, but only the leaves of particular plants. Any old leaf will not do. What my mother didn't know and I certainly didn't either is that 1) the caterpillar had already eaten its last leaf, which is why it was crawling on the ground searching for a place to pupate, and 2) if it had been hungry, it would have starved to death rather than eat the grass and random collection of leaves I had supplied it with. It's not that caterpillars are stubborn, but they are physiologically locked into eating the plants they have coevolved with.

In a nutshell, plants do their best to discourage plant eaters, including insects, from feeding on them by deploying both physical and chemical defenses against herbivores. Leaf toughness, sticky trichomes, and sharp thorns and spines are typical physical defenses, while bitter or toxic phytochemicals such as nicotine, tannins, cucurbitacins, cardiac glycosides, glucosinolates, and pyrethrins are common chemical defenses. Most plant lineages have developed their own unique cocktail of chemical defenses. Collectively, these defenses are very effective and keep most of the insects of the world from eating most of the plants of the world.

But over the eons, insect herbivores have been able to develop specialized enzymes that store or excrete particular phytochemical defenses in particular plants and behavioral adaptations that minimize the insect's exposure to these compounds. These adaptations are so effective that the insect species that have evolved them can circumvent the defenses of specific plants—but only plants that have the chemical that matches the insect's specialized adaptations. So by specializing on one type of chemical defense produced by one plant lineage, the insect species becomes physiologically locked into being able to eat only plants in that lineage. In short, dietary fussiness on the part of caterpillars enables them to use particular plants for growth and reproduction (the good news), but it also constrains them to depending on just a few types of plants for food (the bad news). This is why most North American insects are unable to use most of the plants that have been introduced from other continents: they have not been exposed to those plants for the thousands or even millions of generations required for them to evolve the ability to eat those plants. Plant choice matters, folks; if we don't stick to the plants our insects coevolved with, we will lose those insects and the ecosystems they run.

I have planted all the keystone plants in my area, have reduced my lawn by more than half, have a small pond, and have added a diverse pollinator garden. I've also changed to yellow outside lights. I've got some bees but no butterflies this year! Last year I had many more bees and numerous butterfly species! I'd appreciate any feedback about what could be going on.

Your yard sounds like a paradise! Butterfly populations fluctuate. Some years are good, others are not so good. Sometimes the predators take a heavy toll and it takes a year or two to rebound. Sometimes the weather hammers overwintering butterflies. Be patient. They will be back—unless your neighbor or township is fogging for mosquitoes.

What do you mean when you say there are "false host plant records" in the literature?

People can record a host plant incorrectly in several ways. A true host plant can support larval development from hatching until pupation. Some caterpillar species become more catholic in their food choices just before they finish growing, but they would never have been able to use those plants when they were young larvae. Another way hosts can be recorded incorrectly is when someone finds a cocoon on a particular plant. For example, cecropia moth is recorded as eating ginkgo. That's a false record—someone found a cecropia cocoon on a ginkgo tree and concluded that the caterpillar had been eating ginkgo leaves. Caterpillars usually leave their true host plants and crawl a good distance before they spin their cocoons. They are trying to distance themselves from their host plant because parasitoids often search for larvae and pupae by first locating the host plant. There is a cecropia cocoon on my back porch, but that doesn't mean the cecropia was eating my back porch! Tent caterpillars will leave their black cherry hosts to spin a cocoon and are often found on other plants when they are searching for a pupation site. But those plants are not true hosts.

How do I encourage fritillary butterflies?

Most fritillary butterflies are host plant specialists on various violet species. Females lay eggs in late summer, and the larvae hatch just before violets die back for winter. Surprisingly, the larvae overwinter without feeding. So to have fritillaries, you need lots of violets that are unmowed and untrampled. You also need nectar sources for the adults. The males come out first (early June in southeast Pennsylvania), with females emerging at least three weeks later. Mating occurs, and then females feed on nectar for a few weeks until they start to lay eggs. So summer-blooming plants to be used as nectar sources near your violets are critically important.

How much are outdoor lights hurting insects?

Light pollution is of one of the major causes of global insect decline. Nearly everywhere, we humans light up night skies with white light bulbs that emit cool wavelengths of light (between 3100 and 4500 Kelvin). Unfortunately, these are precisely the wavelengths that most nocturnal insects are attracted to. No one really knows why insects are attracted to artificial lights. The most popular explanation is that night-flying insects orient using the moon as a guide, keeping their flight trajectory perpendicular to the light reflected by the moon. Since the moon is so far away, this angle never changes. But with nearby artificial light sources, the only way they can fly at a right angle to the light is to fly in circles around the bulb. Their flight circles tend to decrease in radiuses until the insects crash into the light or come to rest on a surface near it.

Mortality is high for insects attracted to lights. Once an insect arrives at a light, things rarely turn out well. Many species of moths, for example, fly around and around the light until they literally die of exhaustion. Incandescent lights get very hot and insects die when they fly into the light repeatedly, eventually getting incinerated. All of this activity near a hot light causes insects to die from dehydration as well. Creatures that eat insects learn quickly which local lights attract them, and they dine on the insects gathered at those lights both during the night and early the next morning when many of those insects are still hanging around on the building near the light. These predators include bats, skunks, daddy longlegs, praying mantids, ants, mantisflies, carabid beetles, toads, spiders, and bittacid hanging scorpionflies. Even if an insect escapes all of those dangers, it will likely be blinded by the bright light, which, in turn, will lead to an early death. More subtle but just as harmful to insect fitness is that light pollution disrupts insect circadian rhythms, triggers inappropriate egg laying, and keeps attracted insects from foraging, mating, and reproducing.

When you consider all of the lights that are turned on every night of the year, it's no wonder light pollution is being blamed for insect declines. In this case, however, there is good news. Out of all of the causes of insect declines, light pollution is perhaps the easiest to fix. We can end the carnage on our properties by flicking a switch. Most of us turn lights on at night out of cultural habit, not because of any real security risk. But if security is an issue, we can install motion detectors that activate our lights only when they detect movement. The really good news is that there is an effective solution even for those who don't want to turn their lights off or use a motion detector: replace the white bulbs with yellow bulbs. For some reason, yellow, or so-called warm light wavelengths (those near 2000K), are far less attractive to nocturnal insects than are cool wavelengths. By simply switching the type of light bulb we use outdoors, we can nearly eliminate this huge source of insect mortality. I wish all conservation challenges were this easy to fix.

More good news—and this is starting to happen. Recently, my wife, Cindy, and I went to a campground in a state park in Utah where the bulbs over the bathroom doors were yellow! To my joy (and selfish disappointment—I love looking for moths at campground lights), not a single insect came to those lights that night. I was impressed by how well this simple act of conservation worked.

Which light wavelengths are best if you want to avoid killing insects?

We humans can see light in the wavelength range of 400 to 800 nanometers (nm), but insects can see—in fact, they prefer—light in the ultraviolet range of wavelengths between 320 and 400 nm. LED lights are preferable because they use much less electricity, do not emanate heat, and usually are not as bright as incandescent lights. But most LEDs come in wavelengths ranging from 350 to 700 nm, so just using an LED is not good enough; you have to get one emitting wavelengths at the higher end of the scale. What makes things a bit confusing is LED colors are reported in degrees Kelvin (K): lower numbers (2500–3000K) are warmer, more yellowish lights, so look for these

If we replace the white bulbs in our outdoor lights with yellow bulbs, we could save millions of insects every night.

when buying LEDs that won't attract insects. As you near 6000K, the light produced is more blue/purple and much more attractive to insects. One word of caution: Some LEDs emit much higher wavelengths (4000K and higher) but are painted yellow. These will still attract insects, so be sure to avoid them.

Am I right to think that we should be using red lights when we have outdoor lighting? Insects and moths and birds apparently can't see red at all, which means their diurnal cycles are not disturbed by it, cycles including feeding and mating and rest. Even plants are unaffected, as opposed to what happens with light more along the blue scale of the spectrum.

Because red light does not dilate the pupils as much as other light, we retain our night vision when looking at areas in the dark that are outside the immediate light source, which means it is safer if the concern is bad actors lurking in parking lots just outside the circle of light. Finally, everyone sleeps better, because red light also doesn't disturb our sleep patterns. So why do you and others recommend yellow light?

You're right about red wavelengths. I talk about yellow bulbs because they do not attract nocturnal insects (or at least are far less attractive) and they are less objectionable to humans than red lights. I hesitate to recommend something that I think will be rejected out of hand, but maybe I should get over that. Thanks for the advice.

Pollinators/Native Bees

I have heard that a third of our crops depend on pollinators. Is that true?

I shudder every time I hear the media justify concern for bees by claiming they pollinate a third of our crops. May Berenbaum, chair of the Department of Entomology at the University of Illinois and one of the (if not *the*) most respected entomologists in the world, recently wondered where that statistic came from. After considerable sleuthing (and Dr. Berenbaum is the best sleuther around), she was unable to track down any study that supports the one-third claim. So she made her own estimates.

By her calculations, only one-seventh of our crops depend on animal pollination, and in the typical American diet, which is based heavily on corn and wheat (both wind-pollinated crops), only one-twelfth of our crops are pollinator dependent. What bothers me, though, is not that the media has mistakenly inflated the importance of pollinators to agriculture, but that they believe the *only* value of species such as pollinators comes from how they directly serve humans. That level of anthropocentrism misses the real impact of pollinators—not just on humans, but on most other earthlings as well.

Regardless of their importance to our crops, pollinators are essential to life as we know it: they pollinate 80 percent of all plants and 90 percent of all flowering plants—the very plants that turn the sun's energy into the food that supports animal life on terrestrial Earth. If we recklessly landscape in ways that do not support pollinators, we will lose the species that enable most of our plant species to reproduce—not an option if we humans plan to continue inhabiting planet Earth.

In your writings, you refer to many species of bees that can rear their larvae only on the pollen of particular plants. Do these specialists consume pollen and nectar from the flowers of other native plants to survive until their host blooms, but then use the pollen of their host plant exclusively for their larvae? Or are they in their adult stage only while their host plants are blooming?

A little bit of both. A lot of synchrony exists between when specialist bees emerge as adults and when their required host plant blooms. But the synchrony is rarely perfect, so if a bee emerges before its host blooms, it will nectar on any flower that matches the morphology of its mouthparts (primarily length of tongue) and wait until the pollen it needs to reproduce is available. This is quite common for male bees because they emerge before the females. If the required host does not bloom that season or if it has been destroyed by "development," the bee will never reproduce and its population will decline. This is why seeing a bee sip nectar at a flower does not mean the bee is getting all that it needs from that flower. Its critical resource is pollen from the plant it specialized on over evolutionary time, something non-native plants cannot supply for any North American specialist bee. And there are more than 1300 species of specialists in North America!

Is mixing native plants that would not normally occur together helpful or harmful? For example, would a stand of common milkweed mixed with little bluestem be disruptive to a monarch, to the milkweed, or to the grass? Would it affect pollination?

It is true that particular plants often grow in "associations" such that when you find one member of the association, you often find others. I believe this has more to do with members of the association requiring similar environmental conditions such as soil type, acidity, moisture level, and so on, rather than the species benefiting from each other. Flower visitors are quite opportunistic and take advantage of the pollen and nectar they are seeking wherever they can find it. I'm sure no one has studied whether a mixed planting of milkweed and little bluestem is disruptive to a monarch, but my prediction is that it would not disrupt monarchs or any other flower visitors one bit.

Can you refer me to the best research that supports pollinator corridors made up of small plots of land?

Several studies, when taken together, clearly demonstrate the benefits of small home pollinator gardens to local pollinators. For example, in a 2017 Michigan experiment, Maria-Carolina Simao and colleagues found a linear relationship between the number of flowers present and the abundance and diversity of native bees in small gardens. Moreover, this relationship was stronger in the second year of the study, suggesting that small urban gardens actually help build pollinator populations over time. Similar results have been found using citizen scientists in Sweden. A large meta-analysis conducted by Philip Donkersley and others examined the results of 31 different studies and found that small pollinator gardens always benefit pollinator communities, but they are more effective when they are densely aggregated throughout urban landscapes. These results should be more than enough to encourage homeowners to install pollinator gardens—even small ones.

We read how non-native flowers are like bubblegum and are not really nutritious. I was wondering what you think about including these "alien" plants as pollinator plants?

Non-native plants can provide nectar for some pollinators and pollen for a few more. Remember that honey bees are also non-natives and have no trouble using many of these plants. But the species that benefit from non-natives are generalist pollinators that are adapted to using the pollen and nectar of many species. This is why allowing white clover, a non-native, in your lawn helps honey bees and some bumble bees. The species that do not benefit are the specialist bees that can reproduce only on the pollen of particular native plants. The ideal pollinator garden has lots of plants that specialists need (such as goldenrods, asters, perennial sunflowers, evening primrose, violets, and so on) because the generalist species can use those plants as well. The part about non-native plants having nectar that is like bubblegum is an urban legend. No truth to it. What's worse is that that statement is often wrongly attributed to me!

Annie White at the University of Vermont found that when two species of *Lobelia* are hybridized, the nectar has 20 percent less energy. Is nectar with a lower level of sugar less valuable because the pollinators have to work harder? I guess I'm thinking not all nectars are created equal?

That's right. Pollen differs in nutritional quality, and nectar can differ in the amount of sugar content (energy) it provides. But native plants differ a lot in pollen and nectar nutrition too, so it is risky to compare what a *Lobelia* species contributes to what a butterfly bush or any other plant does. Not only does nectar vary in sugar content from species to species, but it also varies during the day. If water evaporates from nectar, the sugar content of what remains is higher.

I am wondering how I can help a pollinator species that is in trouble but that could thrive if it could find food every few houses.

It is important to recognize that refueling on nectar is one challenge pollinators face, but for many bees, the real challenge is finding enough of the pollen required to reproduce. For example, the mining bee, *Andrena phaceliae*, can reproduce only on pollen from plants in the genus *Phacelia*. Only one plant species, *Phacelia dubia*, grows in eastern North America. So the bee is in trouble wherever *P. dubia* has been removed by landscaping practices. It is doubtful that the bee could forage more than a few hundred yards for new plants. That said, a few years after we planted *P. dubia* at our house in southeast Pennsylvania, the mining bee showed up, which suggests that we *can* restore populations by replacing the plants they need. The same story can be written for *A. asteroides*, which requires native asters; for *A. erigeniae*, which requires spring beauties; for *Colletes albescens*, which requires false indigo; and for dozens of other specialist bees. These are just a few examples, but native bee declines are a serious problem across the country. A recent study by the Center for Biological Diversity found that more than half of the 3600 species of native bees in North America are declining, and 25 percent of them are imperiled. What this means is that planting diverse pollinator gardens throughout a neighborhood—focusing on the plants specialist bee species require to reproduce—will likely help many species of bees increase their population sizes and make their futures a little brighter.

Crape myrtles have been planted in this country for a long time. Have they evolved yet to the point where they can support pollinators and caterpillars? A fellow master gardener said he has bees on his crape myrtles throughout the season.

Ah, we do love those beautiful crape myrtles, and we work hard to justify them in our landscapes. No, crape myrtles have not evolved to help native bees, even though they have been planted in our country more than a century. Your friend who sees bees on his crape myrtle is not fibbing; he sees non-native honey bees. If honey bees were the only bees we needed to sustain our ecosystems, then crape myrtles would be doing half of their job. (The other half is to support the caterpillars that birds need to reproduce. Let me know if you find even one caterpillar on your crape myrtle!) But we *can* justify using crape myrtles as long as we don't overuse them to the point where they dominate the landscape, which they do in so many southern cities.

Many species of native bees in this country are in trouble because we have removed the plants they need to reproduce and have replaced them with plants that don't functionally flower at all (such as lawn and bamboo) or with Asian plants like crape myrtle. Crape myrtle does nothing for any specialist species, and, judging from the number of native bees I see on crape myrtle, very little for our generalist species either. So a few crape myrtles as decorations in our landscapes is OK, but only when they are accompanied by the oaks, goldenrods, asters, sunflowers, evening primroses, and other native plants that support thousands of species, not just the honey bee.

Hundreds of native bees are nesting in the ground in my backyard, and they are swarming everywhere. I'm afraid to mow. What should I do?

I'm guessing that those are colletid ground-nesting native bees. They're totally harmless—you can't *make* them sting you. They are in your yard because your soil is ideal for tunneling and there is good forage nearby. You can't relocate them without killing them, but you can ignore them. You can also cut the grass, no problem. If the wheels of your mower run over a burrow, the female bee will dig it out again. The very best route is to ignore them and appreciate the wonderful gift of pollination your yard is supplying for you and your neighbors.

Why doesn't my Virginia creeper make flowers and berries?

This could be happening for one of two reasons: it does not have enough sun or it is a very young plant and hasn't reached maturity yet. I'd bet it's the former. Plants get their energy from the sun; they can live and sometimes even grow for long periods in shade, but they never accumulate enough energy to produce flowers. No flowers, no fruit.

I'm surprised by how few early blooming natives we have and the huge volume of non-native plants that seem to be sustaining pollinators early in the season. Doesn't that make invasive plants essential to our pollinators?

You are giving our natives short shrift. A sequence of native early bloomers sustained all of our early bees for millions of years before we brought over the invasives. Native willows appear first, then spicebush, followed by at least a dozen species of spring ephemerals, followed by redbud, American plum, amelanchier, silverbell, red buckeye, coral honeysuckle, spring beauties, bluets, blackhaw viburnum, and then black cherry, arrowwood viburnum, phacelia, and on into our summer bloomers. Even oak and hickory catkins provide a lot of pollen for early bees, despite the fact that they are wind pollinated. These plants aren't where you walk because the invasives have pushed them out and/or white-tailed deer have eaten them. An important point is that only generalist bees can use the invasives. None of our 1300 species of specialist bees can use them, so the invasives are not good replacements. You can help both our generalist *and* specialist bees using native plants by removing any invasives in your yard, encouraging your neighbors to do the same, and replacing the non-natives with the early blooming natives.

In your book *Nature's Best Hope*, you indicate that credit given to butterflies for being great pollinators is undeserved, which I took to mean

either they are not pollinators or their pollination activity is insignificant. What characteristic do they possess that makes them important?

Every cog in nature's wheel is important! Butterflies are part of the species complex that runs the ecosystems that support us. Butterflies get a lot of attention from humans because they are pretty, but ecologically speaking, moths are much more important. Many moths are good pollinators at night, but most are important parts of the food webs that support vertebrates, particularly birds. For every single butterfly species, there are 19 species of moths! The more species in an ecosystem, the more productive and stable it is. In that sense, butterflies do contribute to ecosystem productivity and stability.

I will offer one caveat to the generality that butterflies are not good pollinators. Apparently, swallowtails, particularly large tiger swallowtails, do play an important role in pollinating native azaleas and rhododendrons. For a plant to reproduce, a pollinator—usually an insect—has to spread the pollen from the male anther to the female stigma. But in azaleas and rhododendrons, the distance between these two structures is so large that it is unlikely that a bee or other small pollinator will come into contact with both anther and stigma during a visit.

Biologist Mary Jane Epps was interested in how azalea flower structure affects its pollination. She and colleagues Suzanne Allison and Lorne Wolfe (2015) discovered that only butterflies seem to contact both anther and stigma when visiting azaleas, so they tested whether butterflies are actually the primary pollinators of these plants. They set up three different treatments among the azaleas: excluding all pollinators, excluding butterflies but allowing smaller pollinators like bees or flies access to the flowers, and leaving the flowers open to every insect and tracking the pollination process. They found very low rates of pollination when all pollinators were excluded and when only butterflies were excluded. But when the azaleas were exposed to all pollinators, including butterflies, pollination rates were 10 times higher. The flowers of these plants provide no purchase for large butterflies to rest on while nectaring. This forces the butterflies to flutter in front of the flower while they sip nectar. Those wing flutters are strong enough to blow the pollen from the anthers to the ovaries in the stigmas of the flowers.

The University of Minnesota is pushing the "bee lawn" concept. We were onboard until we read your books. A bee lawn is nothing but a mix of alien plants that bees will visit—Dutch clover, creeping thyme, ground plum, and fescue. Do we plant it, or do we resist and let our natives do the heavy lifting?

This sounds like a mix designed for European honey bees. But our native bees also need to eat and reproduce, and Minnesota has several hundred species of native bees. All bees need nectar for energy and pollen for reproduction. Most species are not too fussy about where they get their nectar—as long as their tongues can reach the nectar load of a flower, they will try to use it, native or not. But pollen requirements are often much more specific. Many species of native bees can rear young only on the pollen of particular plant genera. Perennial sunflower species, asters, native willows, goldenrods, evening primrose, and violets all support lots of specialist bees. Those six plant genera alone could support more than 75 species of native bees in Minnesota. That said, if your choice is between a bee lawn designed for honey bees and a traditional lawn designed for beauty only, I would certainly go with the bee lawn.

In light of the insect apocalypse, I would like to measure insect biomass as well as diversity in my yard, if only to have overall measurements to compare with a standard lawn. How do I do this without impacting the tiny ecosystem?

Measuring insect numbers and diversity has always been a challenge because insects occupy so many niches. To say you want to sample insects using one method is like saying you want to sample all mammals using one method. And to do it nondestructively (which I commend) is even tougher. I recommend sampling the two most important groups (pollinators and caterpillars), and using them as indices of the how the rest are doing. But before that, make a complete inventory of the plants in your yard. You win or lose with plant choice. Then sample the insects that eat and pollinate those plants. All the predators, parasitoids, and detritivores will be there if the herbivores and pollinators are there. I figure this out photographically, but that requires an investment in close-up imaging gear. Or you could simply count numbers and assume diversity will be linked. That is certainly the easiest thing to do.

How do we know if we are making progress in our yards and environs? Are the levels of insect noise (more is better), insect and bird activity, and quality of soil good measures?

All of those could be used to measure success, but I think the best measurement is the number of birds that nest in your yard or forage for food for their nestlings in your yard, even if the nest is in a neighbor's yard. Nesting birds are indicative of a great measure of success; they're in your yard because they're finding enough food there to keep their nestlings fed. For most bird species, that means thousands and thousands of caterpillars. You could also look for caterpillar feeding scars on the leaves of your plants. A plant with no holes in its leaves is not supporting the insects that feed the birds. Finally, the loudness (is that a word?) of the katydids singing at night in your yard is a relevant measure from the Midwest to the East.

I know many bees require the pollen of certain plants to reproduce, but do plants require certain bees?

Many plants have developed a dependence on particular bee morphologies, such as tongues of certain lengths or particular pollination behaviors such as buzz pollination (when bees use their wings to create vibrations that eject pollen). Over the eons, the flowers of such plants have evolved to attract the bee species that are best at pollinating them. The traits we find so beautiful— the petal size, shape, and color, including brilliant colors visible only in the ultraviolet spectrum—all have the single purpose of attracting the bee species most likely to transfer pollen successfully from the male flower parts to the female flower parts. Specialized relationships tie plants to certain bees and tie bees to certain plants. The future of pollination depends on keeping these relationships intact.

Can I water areas where native bees are nesting?

I'm glad you asked this question so I can emphatically answer *NO, do not water your ground-nesting bees!* They have a hard enough time surviving natural rainstorms. Watering just turns their soil homes into mud, and a lot of water will collapse their tunnels altogether.

How do I clean out my bee hotels?

Line the holes you drill in wood with a paper tube, such as a straw. Then, after each use, you can remove the straw and replace it with a new one. For overwintering bees, gently remove the straw in March (if it contains a bee pupa) and put it in a protected area so the bee can emerge. This enables you to clean the hotel and put in new straws for spring bees. Apparently, in nature, bee holes are used only once. You can also purchase bee nest liners of various sizes.

What is the most important thing to remember when planting a pollinator garden?

Two things are equally important. The first is meeting the needs of specialist bees. I have addressed this in several answers in this book, but I stress here that planting the species required by specialist bees also supports generalist bees, because they can also use those plants. The other, and perhaps most challenging, aspect of the perfect pollinator garden is selecting plants that will flower in sequence throughout the season. For example, in southeast Pennsylvania, species such as native willows, redbud, serviceberry, wood poppies, and American plum, as well as numerous spring ephemerals, provide early spring forage. Phacelias, fringe tree, blueberries, silverbell, black cherry, coral honeysuckle, wild geraniums, viburnums, and black locusts bloom in late spring. Bee balm, native roses, black-eyed Susans, various milkweed species, bottlebrush buckeye, winterberry, and Virginia creeper flower in early summer. Buttonbush, sweet pepperbush, evening primrose, Joe Pye weed, ditch daisy, and perennial sunflowers bloom in midsummer. And, finally, various goldenrods, New York ironweed, and native asters flower in fall.

These are some examples that I am aware of because they bloom in sequence in my yard. This list includes a lot of plants, but it's still only a fraction of the species you could have in your yard to create a seasons-long sequence of blooms. If you don't have enough space to plant species that provide blooms all season, you might be able to collaborate with one or two neighbors! Note that many of these sources of pollen and nectar are woody plants. Woodies are often overlooked when designing pollinator gardens. Another important point is that not all plants important for pollinators produce big, showy flowers. Virginia creeper and winterberry (in fact, all of the native hollies), for example, have tiny, inconspicuous flowers that you don't even notice until you see all of the native bees hovering around them. Pollinator gardens should be primarily designed for pollinators; if they are not always award-winning splashes of color for us, well, that's alright.

For the wooden kind of bee hotel, how thick should the untreated blocks be, and what circumference and depth should the few holes in each be? By what time of the year do the bees need the hotels? On the ground or in a higher position? How many feet apart? Do they need anything to keep rain or snow off the entrances?

I wish more people would see bee hotels as potential solutions to declining bee populations. Wooden nesting blocks should be around 9 inches thick. Drill holes all the way through—the ideal hole is 8 inches deep. Most drill bits aren't that long, so you may need to drill in from both ends and hope your holes meet up. The holes should be ¼–½ in. in diameter; if you drill holes with different diameters, you can support different bee species.

Set out your bee hotels by mid-March, and don't put them on the ground; elevate them by at least 12 inches. It's best to keep them dry, so making a small roof over each hotel is ideal. Bee hotels should be small, consisting of only five or six holes each. Space them around your yard, and don't place one next to another. The idea is to separate them, so if a predator, parasitoid, or disease finds one hotel's bees, it will not have found all of the hotels' bees—you don't want to put all of the bees in one basket.

Do honey bees compete with native bees?

With few exceptions, they do, and it's easy to understand why. Honey bees are not part of natural pollinator communities in North America. They are domesticated and are artificially inserted into both agricultural and natural habitats in very large numbers. A single hive can house 30,000 bees, so even a small hobbyist can easily add more than 100,000 bees to a particular area.

Honey bees are extreme generalists that forage for pollen and nectar on a wide variety of plants. There are a small number of native plants that certain native bees favor and honey bees do not use. To keep their hives healthy, beekeepers supplement them with sugar so that even when there are few blooming flowers available, hive numbers remain high.

Let's play with some numbers. Honey bees can forage up to two miles from their hive. With that in mind, picture the foraging range for a single hive as a circle surrounding the hive with a diameter of four miles. That circle encompasses more than 8000 acres of land in which bees from that single hive will be taking pollen and nectar from the flowers that native bees depend on. As of April 2022, there were nearly three million honey bee hives in the continental United States. With no overlap, that means that honey bees are potentially impacting native bees on more than 23 billion acres of land, almost 12 times the area of the entire country. Obviously there is overlap among hives, but the point here is that most or all of the country is impacted by honey bee foraging. Let's assume that 25,000 native bees are living within the foraging range of a single honey bee hive. This is just a guess, but it will give us some perspective. We can safely assume that the populations of those native bees are near the carrying capacity of that area—that is, the populations are close to the size that the flowers in that area can support. Now add 30,000 honey bees exploiting those same flowers. How could there *not* be competition for flower resources on all of the plant species that both honey bees and native bees use? Studies have unequivocally demonstrated high levels of competition between honey bees and native bee populations.

The lesson here is that honey bees enhance pollination in agricultural fields. When hobbyists seek to place hives in parks and preserves—or in any natural area—it negatively impacts the native bees already struggling to make a living in that area. It's ironic that surveys have shown that beekeepers almost universally think they are helping the environment. It's time to reevaluate this perception.

I keep honey bees for fun and maybe one day for profit. We all know about the toll that invasive varroa mites take on the honey bee, but has anyone found evidence that the mite parasitizes native bees? Further, my

hives have to put up with wax moths and small hive beetles. Could these also be affecting the native species? I feel a little guilty about beekeeping now that I've been reading your books.

Varroa mites, wax moths, and small hive beetles are specific to honey bees and don't affect native species.

Do insects that overwinter in plant stems choose only pithy stems?

Native bees that overwinter in stems either specialize on pithy stems such as those found in goldenrod, Joe Pye weed, dogbane, and many other plants, or they prefer woody stems derived from soft-wooded plants such as elderberry and willow. Pithy stem nesters choose only pithy stems, but other bee species choose woody stems.

What do you think of "No Mow May"?

Not much. I think it has caught on so quickly because it's so fun to say! In my view, No Mow May makes no mow sense. Maybe that's too harsh, because I do appreciate the intent, which is that we will not mow our lawns in May to help the pollinators. But, ecologically speaking, that won't work.

First, if we don't mow our lawns, we will still have an abundance of long, cool-season European grasses, which, whether long or short, don't supply anything for pollinators. Only if your lawn is loaded with plants that the lawn industry has vilified as weeds (dandelions, clover, deadnettle, and so on) will not mowing in May provide pollen and nectar for non-native honey bees and a few species of generalist bumble bees. This scenario is unlikely, though, because most lawn fertilizers have a broadleaf herbicide that will kill all of those weeds (and many homeowners don't even know this).

Even if your lawn is loaded with plants that will bloom prolifically if you don't mow in May, it seems like a cruel trick on the pollinators to provide food during May and then take it all away in June when you start mowing again. Solitary bees will only be halfway through their reproductive cycles when you pull the resource rug out from under them. Instead of No Mow May, why not create "no mow zones" in parts of your lawn where you *never* mow or only mow once a year, in early March? In other words, let's turn the good intentions of No Mow May into permanent plantings that will actually help pollinators and all flower visitors throughout the seasons. Small meadows or pocket prairies full of plants favored by specialist bees would be great additions to any landscape with enough sun to support them.

CHAPTER TEN

Concluding Questions

In this book, I have attempted to answer a broad spectrum of the questions I have received over the years and continue to receive nearly every day. But it is hardly a comprehensive collection of these questions. I've had to cut many good questions from this book simply to keep it within bounds of a salable product, and I haven't had the chance to address many others. I once considered offering an award to anyone who could ask me a reasonable question I had not been asked before, but I quickly realized that was a bad idea because I get original and thoughtful questions on a regular basis. I particularly enjoy questions about things that I had not previously considered myself. They force me to think outside of my knowledge box and often outside of my comfort zone. Both are healthy exercises, and I hope they have made me a better educator and scientist. Questioning, after all, is the basis of science. Not only have I acquired new information from many of the questions I hear, I've also learned whether my attempts at educating the public have been effective—what messages people are taking away from my books and lectures and, equally important, what messages they are *not* getting. This book is an attempt to fill in some of the gaps in knowledge and to correct misconceptions that I have created over the years. I have placed the following questions into this final section of the book mostly because they are good, frequently asked questions that didn't fit well into any of the preceding chapters.

I often marvel at how developers, loggers, and construction workers can destroy the environment with a clear conscience. Do you have any insights?

People far wiser than I have also pondered that question, so you are in good company. I believe a couple of factors are at work here. First, humans have an enormous capacity to rationalize their actions, and we make unwise decisions every day and think nothing of it. Have you ever smoked a cigarette, driven over the speed limit, gotten drunk, stayed up all night, or eaten a diet high in meat and low in veggies? It is particularly easy to rationalize our behavior when it delivers short-term gain or pleasure. We have not been genetically programmed to prioritize long-term risks over short-term gains. In the old days, we usually didn't live long enough to suffer the long-term disadvantages of bad decisions, so we consistently chose behaviors that provided short-term advantages.

Next, there is the difficulty in seeing the incremental effects of local environmental destruction. Matt Lee-Ashley, who at this writing serves as the chief of staff for the White House Council on Environmental Quality, understood this when he said, "Evaluating the condition of nature is a bit like watching a leaking pipe. If a person focuses on each drop as it falls to the floor, the leak hardly seems damaging. If he leaves for the day, however, he is likely to come back to a room full of water." Moreover, our Western culture still promotes the falsehood that humans and nature cannot coexist, so to make room for humans, we have no choice but to destroy nature. To rationalize this bizarre notion further, we imagine that because there are limitless amounts of nature somewhere else, destroying this little patch won't really hurt anything.

Author Sinclair Lewis perfectly observed that "it is difficult to get a man to understand something when his salary depends *on his not* understanding it." Our developers, loggers, and construction workers depend on their jobs for obvious reasons, and many times their families have depended on them for generations. It is a big ask to expect them to sacrifice their own well-being and that of their families for the greater good.

Why do you think the public's knowledge of basic ecology is so poor?

The ecological IQ of the general public is very low, not because they are stupid, but because ecology was never part of their education. The result of this omission has been a dangerous cultural ignorance about how to exist long term on a finite planet. Even though the term *ecology* was coined by German zoologist Ernst Haeckel way back in 1866, it was not in common use until the mid-1970s. I never even heard the word until I was senior in college, and I was a biology major! I suspect grade-school education in ecological concepts has improved somewhat since then, but evolution has become politicized and many schools are not willing to endure the flack they receive from some parents if they teach kids about it. This is one reason why there is still a sign in one of my neighbor's lawns declaring that evolution is a lie.

We also tend to value expertise and undervalue general knowledge. An economist, for example, is expected to know a lot about economics but nothing about how the ecosystem in their yard works. Only tree-huggers are supposed to know and care about the environment, as if only tree-huggers need a healthy environment. Missing in this type of education are the ABCs that everyone needs to know to be a contributing citizen on planet Earth. This must change. Few of us were home economics majors, but we're all expected to know how to make our beds and cook a basic meal. Because we all have extraordinary powers to disrupt ecosystems in our everyday lives, we all need to understand how they work and learn how not to disrupt them. Every person on the planet needs to achieve basic competency in the ecological interactions that run the ecosystems that people depend on, as well as the stewardship required to sustain those interactions. With this in mind, I am always happy to answer questions about evolution and ecology from people who are willing to think about these concepts with an open mind.

Why is your research not peer reviewed?

Wherever did you get the idea that my research isn't peer reviewed? I have published my research in top-tier journals including *Conservation Biology* (four articles), *Environmental Entomology* (three articles), *Biological Invasions* (three articles), *Ecosphere* (one article), *Diversity and Distributions* (one article), *Ecology Letters* (one article), *Integrative Zoology* (one article), *Journal of the*

American Water Resources Association (one article), *BioScience* (one article), *Heliyon* (one article), *Biological Conservation* (one article), *HortTechnology* (two articles), *Journal of Insect Conservation* (one article), *Proceedings of the National Academy of Sciences* (one article), *Biotropica* (one article), *Restoration Ecology* (one article), *Nature Communications* (one article), *The Condor* (one article), *Ecological Entomology* (one article), *Northeastern Naturalist* (one article), and *Ecological Restoration* (one article). And more publications are in the pipeline. All of these journals are peer reviewed—that is, reviewed by other scientists, many of whom delight in finding flaws in studies. If you want to read any of these publications, visit my website: HomegrownNationalPark.org.

With so many environmental concerns, the average American feels overwhelmed or hopeless when trying to think about them. How would you rate or where would you place the decline and/or loss of species (especially insects, "the little things that run the world") in a world confronting climate change, drought, pollution, famine, and so on—and why?

That's like asking which is worse: being killed in a car crash or in a plane crash? The end result is the same: you're dead. I would rate the loss of insects right up there with climate change. An intolerably hot and erratic climate will destroy the species that run the ecosystems we depend on, but so will the wanton destruction of those ecosystems and their species by humans who think that healthy natural systems are optional. Both forms of ecosystem destruction will end humans as a species. Many studies have shown that ecosystem stability and productivity is directly related to the number of species in an ecosystem. As we take species away, we weaken the ability of ecosystems to reliably produce our life support—the stuff we call ecosystem services.

What gives you hope? How do you not become discouraged by the task at hand?

Several things are happening right now that give me hope about the future of biodiversity, and thus our own future. One is a rapidly growing awareness that there are two, not just one, environmental crises today. Certainly, climate change and the disasters it is wreaking daily across the globe have demanded our attention from the highest echelons of government right down to the barber's chair. Equally important, though, is the recognition that we also have a biodiversity crisis creating serious consequences from declines in plants, insects, amphibians, reptiles, birds, and mammals. This is why the United Nations met in Montreal in 2022 to do what the UN does: make resolutions. The promising aspects of such meetings are not the toothless resolutions they produce, but the fact that biodiversity is on the agenda at all. Even more promising is that the public is starting to recognize that we as individuals are not powerless when it comes to saving our fellow creatures. And this is the big difference between climate change and the biodiversity crisis.

If I asked my neighbor to do something that would help climate change in a measurable way within a year, she would likely think it a hopeless task. She could switch to an electric vehicle and install solar panels, and those things would be incremental steps in the right direction, but she would not be able to see that her efforts had made a difference. However, if I ask her to do something that would help local biodiversity almost immediately, she would be empowered to enhance her yard's ecosystem as well as her greater local ecosystem—with visible effects.

As I hope this book has illustrated, there is much she could do at little cost that would produce dramatic increases in the ecological functionality of her property in very little time. And the fact that the individual is now empowered to become an effective force in conservation means that millions of people can join this effort immediately. More than 332 million people in the United States alone! That's quite an army, and it fills me with hope for the future.

Another positive change I have seen over the last 20 years is the degree to which municipalities across the nation are embracing the need for more native plants. The demand for native plants now exceeds the supply, and in response, the most basic economic law of supply and demand is stimulating the expansion of the native plant industry. Ordinances requiring the use of natives are becoming more common, and financial incentives to replace lawns with natives and remove invasives have been enacted in California, Minnesota, Missouri, North Carolina, and Pennsylvania and are being considered in other states. Statewide bans of invasive plants are now in place in Delaware, Massachusetts, Oregon, and South Carolina, to name a few. Homeowner associations are starting to accept that landscapes dominated by native plants can be tastefully designed and maintained and that the fear of lowered property values is unfounded. There are even important advances in agricultural spaces, such as the increasing use of pollinator strips, the restoration of hedgerows, and the restoration of native plants in roadside verges. These are all indications that our cultural relationship with landscape plants is changing. We are now recognizing that we can, in fact, coexist with the natural world right where we live, work, farm, and play.

What has encouraged me the most, though, is the speed with which these changes are happening. Changing cultural values and the status symbols that entrench those values is difficult, yet I see it happening all around me and wherever I travel. We are moving toward a threshold where native landscaping becomes the norm rather than the renegade exception. Because peer pressure is the strongest motivator of all, that alone will encourage homeowners, developers, and the nursery industry to accept and participate in this inevitable change.

If you were to give advice about the best thing we could do to support our environment, what would it be?

There are so many good things each one of us can do to support our environment, but the most impactful, wide-ranging, far-reaching, long-lasting thing that one person can do is vote—vote with the future of life on Earth in mind. Never, ever vote for a politician who denies the existence of science, who refuses to support ratification of the 1992 Convention on Biological Diversity (the United States is the only country in the world that has not ratified this convention), and who puts short-term monetary gain over the well-being of the ecosystems that support everyone. Ask your environmental advocacy group (Sierra Club, National Wildlife Federation, or one of the many others) for information about the environmental records of candidates who are campaigning for you to let them determine your future, the future of your children, and, indeed, the future of all the plants and animals who call Earth home. We would not have the crises that confront us today if everyone prioritized environmental issues in the voting booth.

I will end with an important question I have often asked myself. Why did we humans change from cultures that typically lived for millennia within the limits of local resources to cultures that over-exploit and thus destroy the very resources on which they depend?

Understanding the answer to this question is important, for it may show us how to return to a sustainable relationship with the natural world. Humans stopped worshiping nature and started worshiping one God sometime after we harnessed agriculture. Apparently, we mistakenly thought that because we could grow our own food, we no longer needed to rely on the bounty provided by local ecosystems and therefore no longer needed to revere them. But were our new beliefs all that different from those of the ancients? I think not.

Most modern religions share three basic tenets:

- God created man and beast and all the fishes in the sea.
- God laid down basic rules (in some cases, commandments) that we must follow to remain in His good graces. If we follow these rules, He is a benevolent God who will grant us eternal life. If we don't follow His rules, we forfeit eternal life and, in some religions, God is vengeful and will punish us.
- God is forgiving when we seek forgiveness.

Let's examine each of these ideas as if we had personified and then deified nature. We will call her Mother Nature.

Did Mother Nature create humans? Based on an increasingly complete fossil record, scientists long ago reached consensus that, yes, humans (and all other life forms) are products of natural selection and evolved among nature's ecosystems from a long lineage of hominids that first appeared about three million years ago in Africa.

Are there basic rules we must follow to remain in Mother Nature's good graces? Absolutely! If we over-exploit the ecosystems that provide our life support—that is, the products from nature called ecosystem services—those ecosystems will collapse and we will lose those services. If we harvest fish faster than those fish can reproduce, the fishery collapses (to wit: Atlantic cod, Alaska's halibut, snow and red king crab, Pacific salmon, bluefin tuna, orange roughy, and many others). If we shoot all of the passenger pigeons, dodos, Steller's sea cows, great auks, Alaskan curlews, Carolina parakeets, American bison, and other animals, those creatures either go globally extinct or become functionally extinct and no longer provide us with meat. If we kill off our pollinators, as has happened in many areas of China and is rapidly happening in the United States, we lose pollination services. If we chop down Earth's tropical forests, we lose the lungs and climate control of the planet. If we pollute the atmosphere with carbon dioxide, we suffer the ravages of climate change. If we pave over our watersheds, we get washed away in floods. If we remove more water from our aquifers and rivers than can be replenished by rain, our rivers no longer reach the sea, the Great Salt Lake becomes a mudflat, the Aral Sea dries up entirely, and our aquifers provide sand rather than water. You get the picture! Mother Nature, not humans, is in control on this planet, and if we don't play by her rules, she will smite us, as it were. The ancients understood this and lived as best they could within the constraints of nature. And when they did not, their societies collapsed.

What about the promise of eternal life for believers? The deification of nature covers that as well. Our genes enjoy an afterlife in future generations. Our kids, grandkids, great-grandkids, cousins, nieces, nephews, and so on, quite literally carry our genes forward, as long as our lineage respects the laws of nature so that we can reproduce those genes.

Is nature forgiving? Within limits, Mother Nature is extremely resilient and extremely forgiving. Given enough time, she can and does repair herself, even after our worst offenses. But she is not endlessly forgiving. There are limits to her ability to rebuild ecosystems as they once were, and the easiest limit to understand is extinction. Mother Nature will not forgive us for condemning one or more of her creations to extinction. The extinct organism, and its contributions to our life support systems, are lost forever. Keep this in mind when you consider the United Nations' prediction that we are on the brink of forcing one million species to extinction.

Is this rhetoric blasphemous? Many will find it so, but it may behoove us to think more in these terms than in terms of the Judeo-Christian edict to "go forth and conquer the earth," an edict we have used to justify our over-exploitation of Earth's bounty for millennia. If God is nature, then every affront to nature is an affront to God.

What interests me most about these ideas is that they do not require great leaps of faith. In fact, they don't require faith at all! The services provided by nature are all measurable. We don't need to *believe* that a marsh community filters pollutants from water, because *we can measure it*. We don't need to *believe* that an ecosystem housing more species is more stable and productive than one from which many species have been removed, because *we can measure it* and see for ourselves. Humans were created by nature and can thrive within the natural world in the peace and tranquility we seek from formal religions, but only if we play by Mother Nature's rules. I believe the ancients had it right: what sustains us on a day-to-day basis is not an unseen deity residing in a place we call Heaven, but is the natural world right here on Earth. The ancients' beliefs did not come about by accident, coincidence, or luck. They were products of natural selection. Those who treated the natural systems that supported them with care and respect thrived. Those who over-exploited them for short-term personal gain perished. If you are looking for a "law of nature," this is a good candidate.

I'm not asking that anyone forgo formal monotheistic religion or a belief in the afterlife. I *am* suggesting that we all have more reverence, more respect, and more care for the natural systems that determine our quality of life. In my view, this is the only viable path forward.

Resources

References

Adelman, C., and B. L. Schwartz. 2011. *The Midwestern Native Garden: Native Alternatives to Nonnative Flowers and Plants*. Athens, OH: Ohio University Press.

Aoyagi, C. 2021. Fire-Resistant Landscaping. *Flora Magazine*, California Native Plant Society, cnps.org/flora-magazine/fire-resistant-landscaping-23654.

Beadle, D., and S. Lecki. 2012. *Peterson Field Guide to Moths of Northeastern North America*. Boston: Mariner Books.

Beilke, E., and J. O'Keefe. 2023. Bats reduce insect density and defoliation in temperate forests: An exclusion experiment. *Ecology* 104(2).

Blossey, B., V. Nusso, et al. 2021. Residence time determines invasiveness and performance of garlic mustard (*Alliaria petiolata*) in North America. *Ecology Letters* 24: 327–336.

Blossey, B., and D. Hare. 2022. Myths, wishful thinking and accountability in predator conservation and management in the United States. *Frontiers in Conservation Science* 3: 2–11.

Brewer, R. 1961. Comparative notes on the life history of the Carolina chickadee. *The Wilson Bulletin* 73(4): 348–373.

Burghardt K. T., et al. 2010. Non-native plants reduce abundance, richness, and host specialization in lepidopteran communities. *Ecosphere* 1(5): 1–22.

Chagnon, M., et al. 2015. Risks of large-scale use of systemic insecticides to ecosystem functioning and services. *Environmental Science and Pollution Research* 22: 119–134.

Condon, M., et al. 2008. Hidden Neotropical diversity: greater than the sum of its parts. *Science* 320(58878): 928–931.

Conrad, A. O., et al. 2020. Incidence and distribution of resistance in a coast live oak/sudden oak death pathosystem. Proceedings of the seventh sudden oak death science and management symposium: healthy plants in a world with *Phytophthora*. General Technical Report PSW-GTR-268. Albany, CA: US Department of Agriculture, Forest Service, Pacific Southwest Research Station: 13.

Copp, C. 2017. *Mighty Oaks from Little Acorns: The Complete Guide to Growing Oak Trees from Seed*. Scotts Valley, CA: CreateSpace Independent Publishing Platform.

de Kiriline Lawrence, L. 1967. A comparative life-history study of four species of woodpeckers. *Ornithological Monographs* 5.

Diboll, N., and H. Cox. 2023. *The Gardener's Guide to Prairie Plants*. Chicago: University of Chicago Press.

Dickie, I. A., et al. 2009. Ectomycorrhizal fungal communities of oak savanna are distinct from forest communities. *Mycologia* 101(4): 473–483.

Dirzo, R., et al. 2014. Defaunation in the Anthropocene. *Science* 345(6195): 401–406.

Donkersley, P., et al. 2023. A little does a lot: Can small-scale planting for pollinators make a difference? *Agriculture, Ecosystems & Environment* 343.

Epps, M. J., et al. 2015. Reproduction in flame azalea (*Rhododendron calendulaceum*, Ericaceae): A rare case of insect wing pollination. *The American Naturalist* 186(2).

Forbes, A. A., et al. 2018. Quantifying the unquantifiable: why Hymenoptera, not Coleoptera, is the most speciose order. *BMC Ecology* 18: 1–11.

Forister, M. L., et al. 2023. Missing the bigger picture: why insect monitoring programs are limited in their ability to document the effects of habitat loss. *Conservation Letters* 16(3).

Frick, T. B., and D. W. Tallamy. 1996. Density and diversity of nontarget insects killed by suburban electric insect traps. *Entomological News* 107: 77–82.

Ginsberg, H. S., et al. 2021. Why Lyme disease is common in the northern US, but rare in the south: The roles of host choice, host-seeking behavior, and tick density. *PLOS Biology* 19(9): e3001396.

Grandez-Rios, J. M., et al. 2015. The effect of host-plant phylogenetic isolation on species richness, composition and specialization of insect herbivores: a comparison between native and exotic hosts. *PLoS ONE* 10: e0138031.

Hadden, E. 2014. *Hellstrip Gardening: Create a Paradise between the Sidewalk and the Curb*. Portland, OR: Timber Press.

Hajek, A. E. 1996. *Entomophaga maimaiga:* A Fungal Pathogen of Gypsy Moth in the Limelight. In Proceedings of the Cornel Community Conference on Biological Control 42: 1–2.

Hallmann, C. A., et al. 2017. More than 75 percent decline over 27 years in total flying insect biomass in protected areas. *PLoS ONE* 12: e0185809.

Hanula, J. L., et al. 2009. Chinese privet (*Ligustrum sinense*) removal and its effect on native plant communities of riparian forests. *Invasive Plant Science and Management* 2: 292–300.

Hettinger, J. 2023. EPA says three widely used pesticides driving hundreds of endangered species toward extinction. *Missouri Independent*,

missouriindependent.com/2023/07/27/epa-says-three-widely-used-pesticide
s-driving-hundreds-of-endangered-species-toward-extinction.

Hoy, J. 2023. Exposure to imidacloprid and glyphosate: effects
on vertebrates. *GMOScience*, gmoscience.org/2023/08/11/
exposure-to-imidacloprid-and-glyphosate-effects-on-vertebrates.

Janzen, D. H. 1968. Host plants as islands in evolutionary and contemporary time.
The American Naturalist 102(928): 592–595.

Janzen, D. H. 1973. Host plants as islands. II. Competition in evolutionary and contemporary time. *The American Naturalist* 107(958): 786–790.

Janzen, D. H. 1988. Ecological characterization of a Costa Rican dry forest caterpillar
fauna. *Biotropica* 20(2): 120–135.

Kays, J., et al. 2015. *The Woods in Your Backyard: Learning to Create and Enhance
Natural Areas around Your Home*. 2nd ed. Ithaca, NY: Plant and Life Sciences Publishing.

Keilson W., et al. 2018. Roadside habitat impacts insect traffic mortality. *Journal of
Insect Conservation* 22(6): 183–188.

Kelso, L. H. 1939. Food habits of prairie dogs. USDA Circular, Volume 529.

Kennedy, A. 2019. Examining breeding bird diets to improve avian conservation
efforts. Ph.D. diss., University of Delaware.

Kleinschmit, J., and B. Lilliston. 2015. Unknown Benefits, Hidden Costs: Neonicotinoid Seed Coatings, Crop Yields and Pollinators. Institute for Agriculture and
Trade Policy, iatp.org/sites/default/files/2015_08_06_Neonics_BL_JK.pdf.

Klem, Daniel Jr. 2021. *Solid Air: Invisible Killer: Saving Billions of Birds from Windows*.
Surrey, BC: Hancock House Publishers.

Kramer, A. J., et al. 2019. Sourcing native plants to support ecosystem function in
urban and rural restoration contexts. *Restoration Ecology* 27(3): 470–476.

Kruckeberg, A. R., and L. Chalker-Scott. 2019. *Gardening with Native Plants of the
Pacific Northwest*. 3rd ed. Seattle: University of Washington Press.

Lambe, D., and L. Edwards Forkner. 2022. *Now Is the Time for Trees: Make an Impact
by Planting the Earth's Most Valuable Resource*. Portland, OR: Timber Press.

Lawson, Nancy. 2017. *The Humane Gardener: Nurturing a Backyard Habitat for Wildlife*. New York: Princeton Architectural Press.

Lecki, S., and D. Beadle. 2018. *Peterson Field Guide to Moths of Southeastern North
America*. Boston: Mariner Books.

Leopold, D. J. 2005. *Native Plants of the Northeast: A guide for Gardening and Conservation*. Portland, OR: Timber Press.

MacArthur, R. 1955. Fluctuations of animal populations and a measure of community
stability. *Ecology* 36(3): 533–536.

Marra. P., and C. Santella. 2016. *Cat Wars: The Devastating Consequences of a Cuddly Killer*. Princeton, NJ: Princeton University Press.

Martin, S. G. 1971. Polygyny in the bobolink: habitat quality and the adaptive complex. Ph.D. diss., Oregon State University.

Mata, L., et al. 2021. Large ecological benefits of small urban greening actions. bioRxiv, biorxiv.org/content/10.1101/2021.07.23.453468v3.full.pdf.

McCargo, H. 2016. In the Shade: Gardening with Native Plants from the Woodland Understory. Wild Seed Project *Native Gardening* blog, wildseedproject.net/2016/03/in-the-shade-gardening-with-native-plants-from-the-woodland-understory.

Mellichamp, L. 2014. *Native Plants of the Southeast: A Comprehensive Guide to the Best 460 Species for the Garden*. Portland, OR: Timber Press.

Mellichamp, L., and P. Gross. 2020. *The Southeast Native Plant Primer: 225 Plants for an Earth-Friendly Garden*. Portland, OR: Timber Press.

Miller, G. O. 2007. *Landscaping with Native Plants of the Southwest*. McGregor, MN: Voyageur Press.

Miller, G. O. 2013. *Landscaping with Native Plants of Texas*. McGregor, MN: Voyageur Press.

Miller, G. O. 2024. *Native Plant Gardening for Birds, Bees & Butterflies: Rocky Mountains*. Cambridge, MN: Adventure Publications.

Narango, D. L., D. W. Tallamy, and P. P. Marra. 2018. Nonnative plants reduce population growth of an insectivorous bird. In *Proceedings of the National Academy of Sciences* 115(45): 11549–11554.

Narango, D. L., et al. 2020. Few keystone plant genera support the majority of Lepidoptera species. *Nature Communications* 11: 5751.

Paine, R. T. 1969. The Pisaster-Tegula interaction: prey patches, predator food preference, and intertidal community structure. *Ecology* 50(6): 950–961.

Pelton, E. 2023. "Keep Monarchs Wild: Why Captive Rearing Isn't the Way to Help Monarchs," *Xerces* blog, Xerces Society for Invertebrate Conservation, xerces.org/blog/keep-monarchs-wild.

Permaculture. 2022. *New York Times* article: "They Fought the Lawn. And the Lawn Lost." Online forum post, Reddit, reddit.com/r/Permaculture/comments/zlv7w2/new_york_times_article_they_fought_the_lawn_and/?rdt=46569.

Richards, M., D. W. Tallamy, and A. Mitchell. 2018. Introduced plants reduce species interactions. *Biological Invasions* 21: 983–992.

Rieske, L. K. 2001. Influence of symbiotic fungal colonization on oak seedling growth and suitability for insect herbivory. *Environmental Entomology* 30(5): 849–854.

Roach, M. 2022. Your Lawn Questions, Answered. *The New York Times*, nytimes.com/2022/06/29/realestate/lawn-care.html.

Roach, M. 2021. Gardeners, Take Heed: It's a 'Tick-y Year'. *The New York Times*, nytimes.com/2021/06/16/realestate/ticks-gardening-2021.html.

Rosenberg, K. V., et al. 2019. Decline of the North American avifauna. *Science* 366(6461):120–124.

Schulte, L. A., et al. 2017. Prairie strips improve biodiversity and the delivery of multiple ecosystem services from corn–soybean croplands. In *Proceedings of the National Academy of Sciences* 114(42): 11247–11252.

Simao, M., et al. 2018. Experimental small-scale flower patches increase species density but not abundance of small urban bees. *Journal of Applied Ecology* 55(4): 1759–1768.

Southwood, T. R. E., and C. E. J. Kennedy. 1983. Trees as islands. *Oikos* 41(3): 359–371.

Sternberg, G. 2021. *Native Trees of North America: From the Rockies to the Atlantic*. Petersburg, IL: Starhill Forest Arboretum.

Sternberg, G., et al. 2004. *Native Trees for North American Landscapes*. Portland, OR: Timber Press.

Stewart, R. M. 1973. Breeding behavior and life history of the Wilson's Warbler. *The Wilson Bulletin* 85(1): 21–30.

Tallamy, D. W., and K. J. Shropshire. 2009. Ranking lepidopteran use of native versus introduced plants. *Conservation Biology* 23(4): 941–947.

Tallamy, D. W., et al. 2020. Do non-native plants contribute to insect declines? *Ecological Entomology* 46(4): 729–742.

United Nations Food and Agriculture Organization. 2020. In Focus: Taking action on climate change, fao.org/home/en.

Valido, A., et al. 2019. Honeybees disrupt the structure and functionality of plant-pollinator networks. *Scientific Reports* 9(1): 4711.

Van Dijk, T. C., et al. 2013. Macro-invertebrate decline in surface water polluted with imidacloprid. *PLoS ONE* 8: e62374.

Wagner, D. 2005. *Caterpillars of Eastern North America: A Guide to Identification and Natural History*. Princeton, NJ: Princeton University Press.

Wagner, D. 2012. *Owlet Caterpillars of Eastern North America*. Princeton, NJ: Princeton University Press.

Wagner, D. L., and K. J. Todd. 2016. New ecological assessment for the emerald ash borer: a cautionary tale about unvetted host-plant literature. *American Entomologist* 62(1): 26–35.

Wagner, D. L., et al. 2021. Insect decline in the Anthropocene: Death by a thousand cuts. In *Proceedings of the National Academy of Sciences* 118(2): e2023989118.

White, D., et al. 1998. Grizzly bear feeding activity at alpine army cutworm moth aggregation sites in northwest Montana. *Canadian Journal of Zoology* 76(2): 221–227.

Wilson, E. O. 1987. The little things that run the world (the importance of conservation of invertebrates). *Conservation Biology* 1(4): 344–346.

Wilson, E. O. 2016. *Half-Earth: Our Planet's Fight for Life*. New York: Liveright.

Wormser, O. 2022. *Lawns into Meadows: Growing a Regenerative Landscape*. 2nd ed. San Francisco: Stone Pier Press.

Online Resources

The American Chestnut Foundation: TACF.org
BugGuide: BugGuide.net/node/view/15740
California Native Plant Society's Calscape: Calscape.org
Homegrown National Park: HomegrownNationalPark.org
iNaturalist: iNaturalist.org
Monarch Watch: MonarchWatch.org
Mt. Cuba Center cultivar studies: MtCubaCenter.org/Research/trial-garden
National Wildlife Federation's Keystone Plants by Ecoregion:
    nwf.org/Garden-for-Wildlife/About/Native-Plants/keystone-plants-by-ecoregion
National Wildlife Federation's Native Plant Finder: nwf.org/NativePlantFinder
Pollinator Pathway: Pollinator-Pathway.org
Save Our Monarchs: SaveOurMonarchs.org
Seek app by iNaturalist: iNaturalist.org/pages/seek_app
Wild Ones: WildOnes.org
Wild Seed Project: WildSeedProject.net
Xerces Society for Invertebrate Conservation: Xerces.org

Sustainability Initiatives

This information was compiled by Ralph Brueggemann.

California
Cash for Grass Rebate Program (Los Angeles County): dpw.lacounty.
    gov/core-service-areas/water-resources/waterworks-districts/
    grass-replacement-program/
Removing Your Lawn (California Department of Water Resources): water.ca.gov/
    Water-Basics/Conservation-Tips/Removing-Your-Lawn

Minnesota
Backyard habitat resources (Minnesota Department of Natural Resources (MDNR):
    dnr.state.mn.us/privatelandhabitat/backyard-habitat.html
Land and Water Conservation Fund (MDNR): dnr.state.mn.us/aboutdnr/law-
    con/index.html
Minnesota pollinator resources (MDNR): dnr.state.mn.us/Pollinator_
    resources/index.html
Minnesota's Legacy Outdoor Heritage Fund (Minnesota State Legislature): legacy.
    mn.gov/Outdoor-Heritage-Fund

Private land habitat (**MDNR**): dnr.state.mn.us/PrivateLandHabitat/index.html
State program pays homeowners to turn lawn into bee-friendly habitat:
StarTribune.com/Program-Pays-Minnesota-Homeowners-to-Let-
Their-Lawn-go-to-the-Bees/510593382

North Dakota
Outdoor Heritage Fund grant program (North Dakota Industrial Commission): ndic.
nd.gov/research-grant-programs/outdoor-heritage-fund

Pennsylvania
Lawn conversion (Department of Conservation and Natural Resources, DCNR):
dcnr.pa.gov/Conservation/Water/LawnConversion
Pennsylvania Heritage Areas Program (DCNR): dcnr.pa.gov/Communities/
HeritageAreas
Pennsylvania National Heritage Areas Reauthorization Act: congress.gov/
bill/117th-congress/senate-bill/378

Index